International Conference on Sustainable System and Maritime Logistics (ISTSML 2021)

MATEC Web of Conferences Volume 339 (2021)

Batumi, Georgia
24 June 2021

Editors:

Parmen Khvedelidze
Badri Gechbaia
Ketevan Goletiani

ISBN: 978-1-7138-3243-0

Printed from e-media with permission by:

Curran Associates, Inc.
57 Morehouse Lane
Red Hook, NY 12571

Some format issues inherent in the e-media version may also appear in this print version.

For additional information, please contact EDP Sciences – Web of Conferences
at the address below.

EDP Sciences – Web of Conferences
17, Avenue du Hoggar
Parc d'Activité de Courtabœuf
BP 112
F-91944 Les Ulis Cedex A
France

Phone: +33 (0) 1 69 18 75 75
Fax: +33 (0) 1 69 28 84 91

contact-edps@webofconferences.org

Additional copies of this publication are available from:

Curran Associates, Inc.
57 Morehouse Lane
Red Hook, NY 12571 USA
Phone: 845-758-0400
Fax: 845-758-2633
Email: curran@proceedings.com
Web: www.proceedings.com

TABLE OF CONTENTS

Shipping Policy of Agri-Food Products and the Formation of Food Markets in Georgia 1
Kharaishvili Eteri, Gechbaia Badri, Erkomaishvili Gulnaz, Lobzhanidze Manana, Natsvlishvili Ia

Intermodal Transportation as a Quality Improvement Tool in Tourism Industry 15
Devadze Anzor, Chaganava Paata, Kordzaia Irakli, Devadze Lela

Risk Management in Maritime Transport of Goods Through Insurance 23
Mitkov Milen, Radukanov Sergey, Petrova Mariana

Actual Problems for Logistics Management and Strategies of Supply Chain in Georgia 32
Tsetskhladze Leila, Makharadze Nanuli, Chkhaidze Irma, Jabnidze Nato, Baratashvili Nazi

Problems of the Formation of Marketing Communications in the Field of Management of Logistics Systems.. 42
Mamuladze Gela, Katamadze Davit, Katamadze Guliko, Mamuladze Giorgi, Bakuradze Ivane

The Main Challenges of Cruise Tourism and Statistics of Expectations in Georgia 51
Abesadze Nino, Kinkladze Rusudan, Giorgobiani Maia, Robitashvili Natalia, Chitaladze Ketevan

The Role of Sustainable Transport System in the Competitiveness of Agri-Food Products............................. 58
Saghareishvili Mariam

The Role of Maritime Logistics and Supply Chain in Stimulating the Export of Agricultural Products 65
Sagareishvili Nino

Criteria for Evaluating and Selecting Suppliers for Maritime Enterprises 73
Lopatin Artem, Ishchenko Natalia, Filimonova Olena, Rudenko Natalia

Indicators of the Seaport Ecological State of a Depressed Industrial Hub 81
Mayorova Irina, Temerbek Alisa, Nosovckaj Oksana, Ryabchuk Valentyn, Ocheredco Elena

Anti-Corruption Audit as an Imperative of Effective Environmental Management.. 89
Nazarova Karina, Nezhyva Mariia, Hordopolov Volodymyr, Mysiuk Viktoriia

Methodical Approach to the Choice of the Variant of the Shipbuilding Company Restructuring 103
Iefimova Ganna, Labartkava Andrei, Pashchenko Oleksiy

Evaluating Performance and Development Priorities of Port Industry in Ukraine..112
Mihai Natalia, Hryshyna Larysa, Khmarska Irina, Pogoryelova Elena, Hryshyna Nataliia

Cruise Tourism in the Red Sea 121
Budz Olena, Romaniv Oksana, Milinchuk Olga, Yarmolyk Dina

Mathematical Modeling of the Trajectory of a Ship as a Control Object in Local Planning........................... 128
Kupraty Oleksandr

Peculiarities of Financing Investment Infrastructure Projects in Ukraine 141
Savina Nataliia, Kovshun Nataliia, Zhemba Alla, Ventsuryk Alisa, Kl³ukha Oksana

Development of the Logistics and Transport System of Organic Products as a Perspective Direction of Ukraine\'s Environmental Strategy 148
Bogomolova Kateryna, Kralia Viktoriia, Podolska Olha, Sahachko Yuliia, Smihunova Olena

Logistics Models of Modernization of Distribution and Transmission Systems for the Development of Ukraine as an Exporter of Green Energy 156
Kovtun Natalia, Yushchenko Nadiia

Suez Canal Blockage: An Analysis of Legal Impact, Risks and Liabilities to the Global Supply Chain 163
Lee Jade Man-yin, Wong Eugene Yin-cheung

Ships' Energy Efficiency Management: Organizational and Economic Aspect 176
Glavatskhih Victoria, Lapkin Alexander, Dmitrieva Larisa, Khodikova Inna, Golovin Alexander

Financial and Investment Processes in Ensuring the Development of Marine Infrastructure 184
Hryshyna Larysa, Pogoryelova Elena, Karas Petro, Mihai Natalia, Khmarska Irina

Author Index

Preface

The International Conference on Sustainable Transport System and Maritime Logistics (ISTSML 2021) is a peer-reviewed international conference that is a leading scientific forum with international participation of experts in the fields of Transport System and Maritime Logistics. ISTSML 2021 conference was held on June 24, 2021.

Sustainable Transport development can be defined as development that meets the needs of the present without compromising the ability of future generations to meet their own needs and preserves human health and their environment, in the way that economy progression will be continued, and community balance will be met.

Sustainable maritime logistics contributes to the achievement of acceptable environmental performance in the supply chain of maritime transport. For this reason the Sustainable Transport System and Maritime Logistics is a potential frame for using sustainable renewable energy sources, preserving biodiversity and the marine ecosystem.

In this context, the aim of ISTSML 2021 is to provide a platform for researchers, entrepreneurs, higher education teachers and authorities to present their research results and development activities in the following conference topics:

SUSTAINABLE TRANSPORT AND ENVIRONMENTAL ECONOMICS;

MARITIME LOGISTICS AND SUPPLY CHAIN MANAGEMENT;

TRANSPORT SYSTEM ANALYSIS AND ECONOMIC EVALUATION.

Due to the global COVID-19 pandemic, the ISTSML 2021 organizing committee decided to conduct the conference in the form of a HYBRID CONFERENCE on June 24 of 2021 combining a live in person event with a virtual component.

The conference was attended by about 100 leading researchers, economists and scientists from different countries: Hong Kong, Bulgaria, Ukraine, Georgia, Estonia, Indonesia. All articles have been thoroughly reviewed and reviewed to meet the requirements of the International Publication Standard.

Conference Chair

Dr. Prof. Parmen Khvedelidze

Rector of Batumi Navigation Teaching University, Georgia

Statement of Peer review

In submitting conference proceedings to *Web of Conferences*, the editors of the proceedings certify to the Publisher that

1. They adhere to its **Policy on Publishing Integrity** in order to safeguard good scientific practice in publishing.
2. All articles have been subjected to peer review administered by the proceedings editors.
3. Reviews have been conducted by expert referees, who have been requested to provide unbiased and constructive comments aimed, whenever possible, at improving the work.
4. Proceedings editors have taken all reasonable steps to ensure the quality of the materials they publish and their decision to accept or reject a paper for publication has been based only on the merits of the work and the relevance to the journal.

Title, date and place of the conference: International Conference on Sustainable Transport System and Maritime Logistics ISTSML 2021

Batumi, Georgia, June 24, 2021

Proceedings editor(s): Parmen Khvedelidze, Badri Gechbaia and Ketevan Goletiani

Date and editor's signature

June 28, 2021

MATEC Web of Conferences **339**, 01001 (2021) https://doi.org/10.1051/matecconf/202133901001
ISTSML 2021

Shipping policy of agri-food products and the formation of food markets in Georgia

Eteri Kharaishvili[1*], *Badri* Gechbaia[2], *Gulnaz* Erkomaishvili[1], *Manana* Lobzhanidze[1], and *Ia* Natsvlishvili[1]

[1]Ivane Javakhishvilli Tbilisi State University, Faculty of Economics and Bussiness, 1 Chavchavadze avenue, 0179 Tbilisi, Georgia
[2]Batumi Shota Rustaveli State University, Department of Business Administration, Management and Marketing, Ninoshvili/Rustaveli str. 35/32, Batumi, Georgia

Abstract. The article evaluates an importance of shipment in economic growth and the development of food markets. Delays caused in shipment of agri-food products due to the COVID-19 pandemic are analyzed, problems in logistics are identified. It is argued that by identifying challenges in the shipment system and developing effective policy mechanisms, it is possible to create food markets that meet the changing requirements. The qualty of impact-emergency matrix identifies challenges in food shipment policy: weak political support, legislative vacuum on shipping and food market standards, non-compliance of regulatory rules with regulations of other countries, low level of digitalization of shipments, policies to maintain food prices in case of increase in shipment costs, etc. The article concludes that the correlation between food production and its shipment rates is positive, however, insufficient to impact economic growth. Pursuing a transport-stimulating policy will lead to the diversification of the food markets, especially in European markets. Considering the development of the food market and the challenges identified in the field of shipment policy, recommendations have been developed towards policy priorities and directions of the development of food markets.

1 Introduction

Shipping business in any country of the world is characterized by a high rate of development. Transportation and logistics plays significant role in sustanble development. Sustainable development is a development that meets the needs of the present without compromising the ability of future generations to meet their own needs [1]. It is based on three pillars: economic, social, and environmental. Sustainable development for transportation systems face main challenges: non-inclusive development, air pollution, adverse health and social impacts, road safety, growing demand, energy consumption and other resources.

* Corresponding author: eter.kharaishvili@tsu.ge

Transport policy decisions made today will have a huge impact on humans's life security in the future. United Nations Center for Regional Development addresses the need for an integrated approach to manage with social, economic, and environmental issues associated with the transportation and logistics sector. Since 2004, has been promoting Environmentaly Sustanble Transport concept [2]. There is no single definition of this concept, but it is centered on the transportation system and transportation activity that meets social, economic and environmental objectives. Air pollution from the transportation sector, growing vehicle ownership, urbanization trend threetens to undermine human health, the urban environment, economic productivity, and social equity. Sustainable transportation methods considers the means by which transportation takes place in line with the principles of sustainability, it is necessary to integrate three specific areas: economic equity, the environment, and social development [3].

Georgia has historically been a corridor between Europe and Asia in terms of shipment. One of the competitive advantages of Georgia is its transit function, which, together with natural-climatic conditions, intellectual resources and other factors, should ensure the development of business in the country and increase the competitiveness of the economy [4]. Due to its favorable geographical location, Georgia has the opportunity to play a key role of a transit corridor, to integrate its transport and logistics systems into international logistics systems and thus contribute to the overall economic development of the country. Ivestments play key role for the development of transit systems in Georgia and it successful integration into the world economy. Investments in the transport and logistics infrastructure of OECD member and partner states with transit function and sea access (among them Georgia) are highly effective in stimulating the transport and logistics sector itself. Such investments also play a very positive role in the socio-economic development of the country [5].

In ensuring sustainable development "An international integrated intermodal transport and logistics system" endorsed by the United Nations [1] plays the significant role. This system is based on the intermodal network of well designed, maintained and interconnected highways, railways, inland waterways, sea ports, river ports, airports or dry ports. The system combines the advantages of each mode of transport to achieve the balanced integration of economic, social and environmental benefits. Based on modal shift the system optimizes the needs of transporting goods and passengers, minimizes consumption of energy, land and other resources, generates low emissions of, greenhouse gases and ozone depleting substances, minimizes the adverse social impacts arising from transport operations. Shipment plays an important role in economic growth. It is closely linked with the development of the agricultural sector that can make a significant contribution to economic growth and poverty alleviation, especially in developing countries [6]. Researchers argue that structural transformation of the sector is critical to the sustainable production of agri-food products. It is necessary to mobilize resources and to encourage innovation policies in the food supply system [7]. Agrifood supply chains are distinguished from other industrial manufacturing and service supply chains by several characteristics: risk emanating from the biological nature of agrifood supply chains, the role of buffer stocks within the supply chain, the scientific foundation of innovation in production agriculture having shifted from chemistry to biology, cyberspace and information technology influences on agrifood supply chains, the prevalent market structure at the farm gate remains oligopsony, relative market power shifts in agrifood supply chains away from food manufacturers downstream to food retailers, globalization of agriculture and agri-food supply chains [8]. At a global level the global food supply chain is connected to the transport network. During the pandemic this network slowed down, therefore the food supply is affected. The aviation industry was in a state of shock. This created an obstacles to the food supply chain and its growth in the short term [9].

Georgia's agricultural sector has the potential to produce and supply competitive agri-food products to the market [10]. One of the most important challenges for the country in the supply chain of agri-food products is the efficient shipment of agri-food products. It should be noted that the food sector is particularly "sensitive" in the Georgia's economy, as food is consumed by all consumers, including those living below the poverty line. Therefore, it is important to ensure that food prices remain stable. Rising prices on these products will lead to the decrease of purchasing power of the population in the face of unchanged income. Rising food prices often exceed the subsistence level, and it is the price change that is the main challenge for the food security of Georgia [11].

The share of food expenditures in Georgia in the total consumer basket is 43% [12]. The problem is exacerbated by the fact that Georgia is heavily dependent on imported food. Researchers believe that the need for applying food pricing policies is even more acute in the middle-income countries. In case of both import and export of products, as well as for the effective trade at the global level considering price increases or reductions, appropriate policy mechanisms are needed [13].

Food prices have risen significantly and become more volatile in the last decade [14]. The sharp change in the prices has affected almost all types of products. The upward trend in prices is also observed for shipments. According to the National Statistics Office, in May, the producer price index for transport and warehousing services increased by 0.8% compared to the previous month. Prices for transport services in Georgia have increased by 2.9% [15]. Clearly, changes in shipping prices will affect food prices. It is therefore important to prioritize food shipment policies that reduce the impact of shipping costs on food prices.

Agri-food products are distinguished by the wide assortment. Due to the perishing of products it is important to use an optimal delivery system in a short time, so it is necessary to specialize shipping systems and to some extent centralize, consider the seasonality of the logistics system, marketing globalization and more. Efficient shipment of agri-food products, provision of appropriate logistics systems and technologies ares important factors for success for both producers and retailers on the one hand and consumers, on the other and as well as for the development of food markets in general.

The COVID-19 pandemic has created a number of logistics related problems that have had a significant impact on the long value chain of the modern global economy. The pandemic is the greatest challenge of a today's world. It poses a threat to the health of millions of people, causing unprecedented global economic mega-recession and uncertainty.

According to the Organization for Economic Cooperation and Development (OECD) [16], pandemic "lockdown" of the economy directly affects each sector of the economy and causes a reduction of almost one third of GDP in the world's leading economies (G20 countries). Lockdown causes a loss of 2% of annual GDP each month. According to the initial assessment of the World Bank, the pandemic and the lockdown of the leading economies can lead to extreme poverty for more than 60 million people. This fact means leveling of all the results achieved in the recent years for poverty eradication [17]. The real difficulties caused by the pandemic were exacerbated and researchers were forced to worsen their predictions. According to a later and more accurate assessment of the World Bank, The world has did a lot to reduce "extreme poverty in recent decades, but was not on track to reach the goal of ending it by 2030 even before the coronavirus hit. The pandemic could push about 100 million more people into extreme poverty in 2020. And by 2030, up to two-thirds of the world's extreme poor could be living in settings affected by fragility, conflict, and violence" [18]. The pandemic poses a particular threat on poor countries in terms of food security, as they have to import food in the event of a fall in their national currency, declining incomes, and difficulties in importing food into the country due to restrictions on export from that country and supply chain disruption [19]. The set of economic recovery measures should include measures aimed at: saving human lives, protecting the most vulnerable groups, reviving business activity and maintaining jobs, as well as creating a shock-resistant and flexible economy.

The pandemic and the measures taken to address it have created new challenges in the food transport system. Countries are faced with the task to achieve the smooth functioning of food markets and food supply chain to meet the growing demand for food. In order for the pandemic not to become the cause of a catastrophic food crisis, it is necessary to make appropriate decisions in accordance with the newly formed requirements of the entire agri-food production and supply chain, from which food shipment is of utmost important.

Efficient shipment of agri-food products will also play an important role in the development of food markets. Therefore, by identifying the challenges in the transport system and developing effective policy mechanisms, it will be possible to structure the food markets in accordance with the changed requirements.

Based on the above, the aim of the study is to develop recommendations on policy priorities and directions for the formation of food markets based on the identification of agri-food shipment policy challenges.

2 Methods

The research is based on the methods of analysis and synthesis, quantitative and qualitative, groupings, relative and average values, graphical representation, comparison, panel data analysis and others. In-depth interviews with experts were conducted to determine quantitative and qualitative indicators; The Impact-Emergency Matrix identifies barriers to food shipment and policy challenges; Priority schemes set the priorities for an economic policy to stimulate shipment.

Demand for agri-food products is growing worldwide and is gradually becoming global. The quantity, quality and safety of food often depend on the efficiency of food delivery systems. Since the second half of the 20th century, significant investments have been made in transport infrastructure, product storage technologies have been developed, large urban aggregates have been created as a result of global urbanization, and more. In general, the urbanization of the economy has brought about significant changes in the food distribution and shipment system. Demand for the food has grown exponentially and gained global scale. Accordingly, special attention has been paid by researchers to the creation of global freight distribution systems [20].

Transport infrastructure plays an important role in the development of agriculture. Scientists have substantiated the hypothesis that the developed transport system helps to increase productivity in agriculture. A country with better transport can produce more agri-food products [21]. Transport is a key factor that can play a crucial role in improving the economic performance of developing countries.

The role and importance of logistics in agri-food supply and supply chains is vital, which is also confirmed by empirical studies. Researchers place particular emphasis on farms and analyze their activities in the light of growing environmental, social and ethical requirements associated with agri-food supply chains [22].

The COVID-19 pandemic has made significant changes to shipment services. Scientists have assessed the impact of pandemic transport disruptions and new transport service requirements on the agri-food supply chain.

Scientists' analysis reveals that access to agri-food products by sea, rail and freight has generally improved during the pandemic. This is due to the declining demand for shipment of products from other sectors of the economy [23]. The pandemic caused significant disruptions in the food supply chain, which in turn affected the efficient functioning of food markets. The limited functioning of food markets poses a threat to the food security of the population. It is therefore important for the state to develop policy mechanisms that will facilitate the targeted movement of agri-food products towards local and global food markets, respond to challenges in the food supply chain, and reduce the negative effects of the pandemic [24]. The negative impact of COVID-19 for food supply chains is caused less from its impact on primary production or overall food demand than from its disruptive effects on the complex network of actors connecting farm to clients, and the unexpected change in the demand mix. Farm production has been affected by bottlenecks for inputs (most notably labour), processing has been disrupted by labour shortages and shutdowns, some modes of transport have been affected more than others, consumer demand has seen rapid and unprecedented shifts [25].

Shipment plays an important role in the smooth and sustainable functioning of food markets. The functioning and consistency of food markets are influenced by the forms and consistency of public policy coordination. Food market formation policy ensures effective links between producers and consumers that makes food more accessible through fiscal incentives [26]. The use of modern digital systems has a special impact on the value chain of agri-food products, especially on food shipment, as well as on agricultural policy instruments [27]. The share of small farms in the total number of farms in developing countries is high. The volume of products produced by small farms is very sensitive to shipment. Shipments of small volumes of product are often made by inadequate or poorly functioning vehicles with consequently low returns. A well-planned shipment system will enable small entrepreneurs to increase productivity, reduce shipment costs, strengthen social integration and improve the quality of life [28].

Studies have shown that it is important to observe safety measures and take into account the optimal time of shipment when transporting food, while preventive monitoring will significantly reduce the risks. However, these issues should be addressed using regulatory rules [29]. Among the challenges in the field of food shipment and distribution, researchers point to the need to expand the Internet and related infrastructure networks. This, in turn, will lead to the sustainability of logistics [30].

Considering the abovementioned, it is important to have a comprehensive view of the directions of shipment and food market development policy, in particular, the policy regarding transportation and regulation of internal food markets can include several areas, the most important of which are: Producer-oriented policy, which manifests itself in various forms of producer support (e.g. production subsidies); Customer-oriented policy, which involves influencing market operations measures (e.g. control of food stocks and prices); Shipping management policy (e.g. setting minimum and maximum shipping prices, stimulating food shipments); Funding policies in the field of digitization of shipments (e.g. acquisition of modern transport and management technologies, promotion of knowledge sharing);Development, monitoring of food safety and quality standards, etc.

3 Results

Modern studies have examined the existing barriers to the implementation of sustainable transport infrastructure policy and proposed practical recommendations for their elimination, including the formation of integrated transport agencies, policy development to regulate transport demand management, cost control for integrated development, and others [31].

The economic literature evaluates transport policy challenges and development directions. Emphasis is placed on the impact of transport investment and policy impact on economic growth in developing countries, as well as specific challenges for transport interventions are discussed [32].

In countries with developed economies, investment in shipping and improved technologies have led to a continuous reduction in spending on shipping, which has contributed to economic growth and, consequently, the further development of these countries. The current potential of shipping policy in low- and middle-income countries is not sufficient to promote sustainable and inclusive growth. Transport infrastructure is not properly developed, regulations in the transport sector are inadequate, the share of transport costs in total costs for food is high and appropriate policy instruments (subsidies, standards, taxes, incentive programs, etc.) are not applied.

In developing countries, special importance is attached to the economic policy of facilitating shipment, as these countries are characterized by asymmetric transport infrastructure. Scholars agree that the economic policies that facilitate shipment have a positive impact on food exports and consequently, on job growth in food companies. The scientists found that the road improvement program led to a significant increase in the average annual export growth rate of firms (6.4%), followed by an increase in employment to 5.1% [33].

Rising food prices in recent years have been the subject of debate among scientists and politicians. In international food markets, price changes are seen as the main challenge to the food security [34].

Investments in transportation in Georgia can play a pivotal role in the loss-free and efficient movement of cargo from the production of agri-food products to consumers. In many cases, transport investments are inefficient and do not yield the expected results, especially with regard to the development of food markets. Given the scarcity of resources, it is necessary to prioritize transport and infrastructure policies, study the factors influencing the formation of transport policy, identify challenges in the field of shipment and work out ways to overcome them.

Analysis of the current structure of the supply chain in Georgia, evaluation of the role of producers, distributors, intermediaries, storage centers and other participants showed that one of the most important problems for the country is the reduction of delivery time from agri-food production to consumption [11]. At the same time, there are quite competitive food markets in Europe and Asia. Consequently, for any country, as well as for the manufacturer, the main task is to supply a competitive product to the market.

A study of the dynamics of cargo transported by the transport sector in Georgia showed that the volume of shipments did not change significantly in 2016-2020 (see Figure 1).

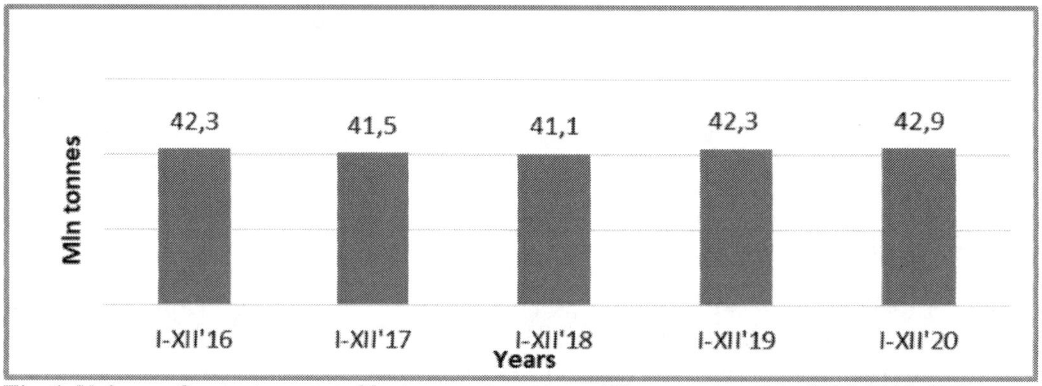

Fig. 1. Volume of cargo transported by transport sectors, million tons (total) in 2016-2020 [35].

As the diagram shows, in 2020, compared to 2016, the number of shipments of cargo increased by only 0.6 million tons. Cargo is transported by road as well as by rail and air (see Figure 2).

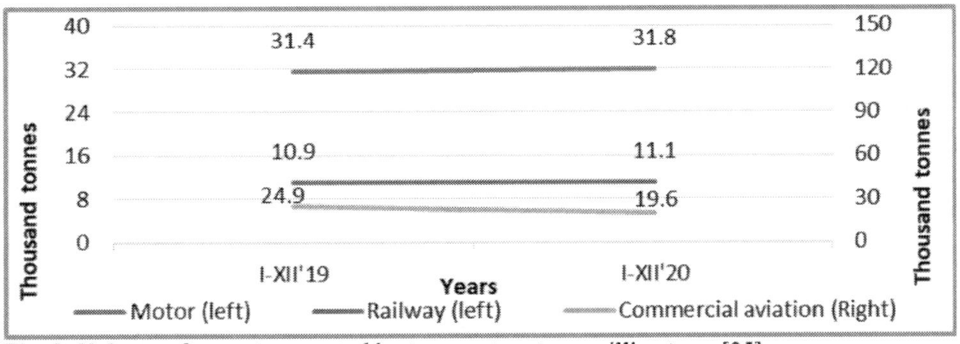

Fig. 2. Volume of cargo transported by transport sectors, million tons [35].

The diagram shows the dynamics of cargo transported by the transport sectors in the country in 2019-2020. The freight rate for both road and rail transport increased in 2020 compared to 2019, while air freight decreased due to the flight ban.

A crucial role in creating a developed transport system can be played the by developing effective mechanisms to stimulate shipping policies.

First of all, it should be noted that cargo shipment in Georgia is regulated by the relevant legislation and regulatory rules [36], however, the current legal framework is vague or imperfect.

Interviews with shipment managers showed that shipping companies have significant problems at this stage, including reduced shipping volumes and increased shipping services. It was also mentioned that in the conditions of the Coronavirus pandemic, the transport companies have to adapt not only to the regulations of Georgia, but also to the regulations of other countries, which makes their operations even more difficult.

A study of carriers has shown that due to the Covid 19 pandemic, some shipping companies in Georgia are completely shut down, some are only working on old orders. The new orders mainly come only for food and medical products. However, these orders are also difficult and delayed, and a limited number of products are supplied to the food markets. The main reason for this is the problems with the drivers (2-week quarantine after delivery, due to which companies are unable to take new orders), as well as increased time on delivery of goods (inspections due to various regulations, delays at the border, disinfection, etc.). Interviews with experts showed that shipping services have risen in price by about 20% due to the named problems [37].

Speaking about the trend of rising shipping prices, experts noted that the rise in prices will also affect food prices. According to experts, decisions must be made by the government so that the increase in transport costs does not have a significant impact on food prices.

The increase in shipping costs is primarily related to food prices and is related to food security for the population, so it is necessary to use government policy tools in this regard. In order to pursue a policy of maintaining food prices, in some cases it is necessary for the government to intervene in private business and impose restrictions.

In Georgia, as mentioned above, the volume of imports significantly exceeds exports. Clearly, under such conditions, the negative effects of the coronavirus impact on food prices are particularly large. In order to reduce the negative effects, 9 main types of products (rice, buckwheat, pasta, oil, flour, wheat, milk powder and sugar and beans) were subsidized in order to maintain stability of food prices in Georgia [38]. The subsidy will be provided in the amount that includes the cost of purchasing products and shipment to Georgia, and in the case of mixed goods, the cost of shipment will be taken into account in the appropriate proportion.

The policy of maintaining food prices in this case is a policy of subsidizing producers. State will compensate companies the amount with which the manufacturer/importer firms have suffered. Like the tax, the benefit received from the subsidy is distributed between buyers and sellers. The nine products mentioned above are primary food products, consumers have an inelastic demand on them and it can be said that consumers will benefit more than manufacturers. Manufacturers are reaping the benefits as they sell their products at a higher price. Food pricing policies will be enacted in the short run, however, due to the economic crisis caused by the pandemic, when a large part of the population is unable to earn a living, pursuing a similar policy by the government will have a positive impact on improving the welfare of the population. This policy is mainly in line with the crisis period and therefore, the policy of maintaining food prices during the pandemic period is justified.

The increase in prices for the shipping service is caused by the extension of the transit terms, as well as the increased delays at the borders. It is well known that many countries have banned the export of certain products, checking this takes time and has a negative impact on shipping terms. Increased demand and reduced quantities of shipments lead to higher prices [37].

Improving the transport system significantly increases market access, which in turn is directly related to the growth of agricultural production [39]. Relevant studies have also found that a 1 percent increase in market access in the country leads to a 0.03 percent increase in GDP per capita [40].

Improving shipment can lead to a relatively even distribution of production across the transport network. Scientists emphasize that the named process affects the clustering of the activities of companies supplying raw materials and food, resulting in increased productivity in terms of economic growth and the development of food markets [41].

We used the Emergency-Impact Quality Matrix to identify challenges in food shipment policy (see Figure 3).

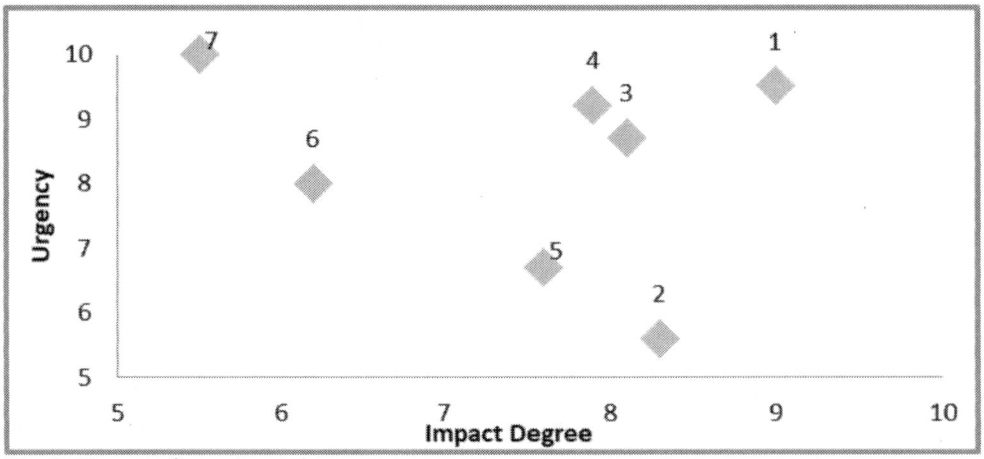

Fig. 3. Degree of emergency-impact.

By the grouping of materials according to the uniformity, the following challenges were identified in the field of food shipment policy in Georgia:

(1) weak political support for food shipment;

(2) a legislative vacuum on food transport and food market standards, as well as a lack of information on existing scant standards;

(3) non-compliance of Georgian transport regulatory rules with the regulations of other countries;

(4) a small number of innovative carrier companies that would be able to quickly adapt and respond to the changing market environment;

(5) Low level of use of digital technology in the shipment system.

(6) Lack of the innovative infrastructure for shipment of food, low level of use of modern technologies;

(7) Food price maintenance policy in the event of an increase in costs of shipment.

It is important to develop economic policy mechanisms to stimulate shipment in order for food markets to function effectively.

We used the "Priority Scheme" to prioritize economic policies to stimulate shipment.

The following main areas were identified as economic policy priorities based on the interviews with the experts (see figure 4).

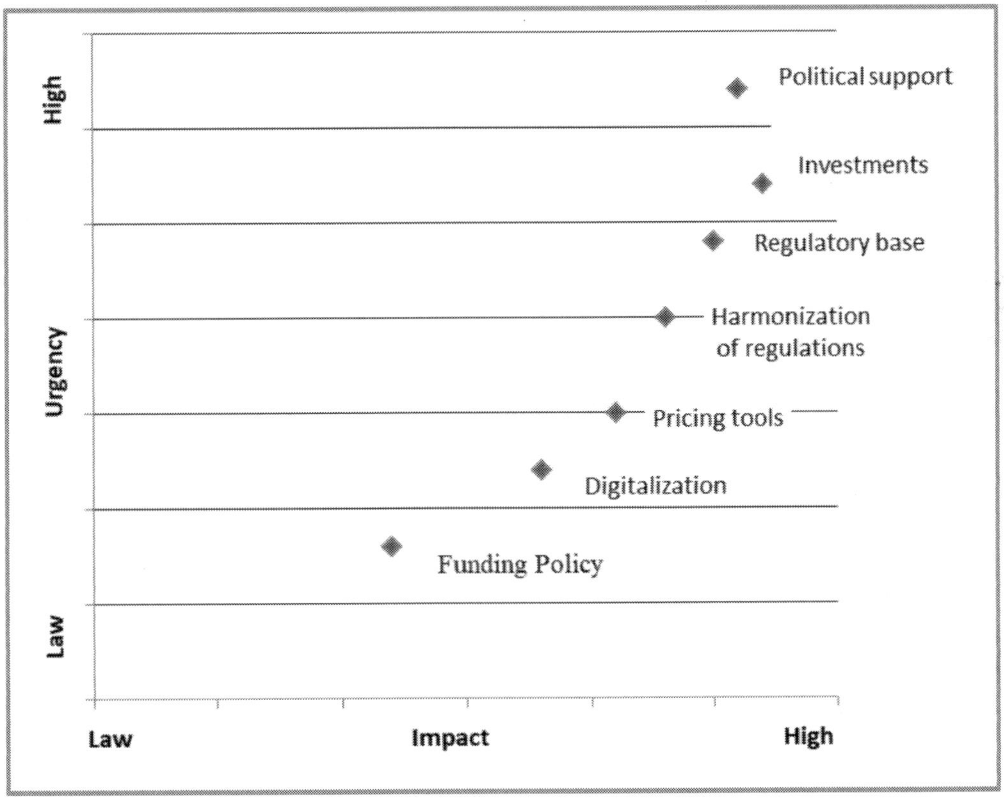

Fig 4. Economic Policy Priorities to Stimulate Shipment of Agri-Food Products

Several key priorities in the field of shipment policy have been identified: 1. The need for political support for food shipment; 2. Investment in the development of shipping infrastructure; 3. Creation / improvement of the shipping regulatory base; 4. Harmonization of the shipping regulatory rules of Georgia with the regulations of other countries; 5. Use of pricing tools in food shipments; 6. Digitalization of the shipment system and the transport corridor; 7. Provide funding policies to maintain a unified chain of shipping and food markets, etc.

Political support is considered to be one of the main priorities of the economic policy to stimulate food shipment. An important direction was the need to continue the investment supporting policy in transport infrastructure, as well as not only to eliminate the ambiguity of the legal framework and create regulatory rules, but also to harmonize shipping regulations with market standards and the legal space of Europe and other countries.

Political interference will affect the supply of food and the development of food markets, for example by making targeted investments in transport infrastructure. In this regard, given the broad objectives of shipping policy, key directions may be to stimulate growth in food production. If shipping costs are low, this will lead to structural changes in the manufacturing and service sectors, increased productivity, the emergence of new industries, creating an agglomeration effect, providing trade incentives, making shipment available, and so on. Clearly, all of the above will help increase the economic opportunities of the low-income population and improve the welfare of the population in general. Achieving the stated goals will depend on the behavior of food firms and households in response to the policy intervention mechanisms.

Farmers in Georgia lack access to food markets, the main reason for this is that they can not use modern shipping systems, modern production technologies, as well as the their demand for targeted vehicles aren't met due to the underdeveloped supply chain. Food shipment will be substantially improved due to the supply chain perfection. In this regard, it is important that the chain of buyer and seller in food delivery is short, which will also lead to the establishment of new relationships between food stocks and the consumers. It is important to reduce food losses, as this figure is 30-40% of the total amount of food, half of which comes from shipment losses. As a result, large quantities of products are not supplied to consumers in the food markets.

The study of the availability of food markets in Georgia showed that the food markets in Tbilisi are unevenly distributed, the unequal distribution in the regions is even more pronounced, therefore one of the ways in improving the shipping system should be developing a food market access policy [42-44]. To ensure greater access to food markets for small farmers, it is essential to develop road systems and appropriate vehicles, especially in the highlands.

In stimulating food market development policies, it is important to increase access to information on markets and targeted vehicles and to close the information vacuum. Creating and delivering comprehensive information packages to beneficiaries, as well as increasing the frequency of deliveries, will contribute to the efficiency of food shipments and the expansion of food markets.

Priorities for price incentives include subsidies or taxes that may affect shipment and shipping behavior, such as food subsidies, fuel tax subsidies, and more.

Pursuing a shipment-stimulating policy will lead to the diversification of food markets, especially in European markets. The correlation between food production and its shipment rates is positive, however, insufficient to impact economic growth. It is important for economic growth to use effective mechanisms to stimulate shipping. The selection of transport policy mechanisms should be made in accordance with the development perspectives of the food market.

4 Conclusions

Shipment infrastructure creates conditions for the formation of the local and national food markets. On the one hand, the effective work of agri-food companies depends on the shipment, which is directly related to its viability in the market, and, on the other hand, the market itself involves the exchange of goods and services, which is impossible without shipment.

Georgian transport companies, which do not have the appropriate technical base, can not compete with international transport companies. Because of this, companies are mainly engaged in domestic shipment;

Georgia does not have a body responsible for the food and transport policy in general, political support is weak, relevant legislation is vague and most of them are incompatible with the European legislative frameworks;

Compared to competitors, Georgia has high shipping rates for sea, land and rail, including terminal services; At the same time, insufficient development of shipment related technological systems leads to a high share of transport costs in total costs. The named factors lead to rising food prices and the loss of the food markets;

No government strategy has been implemented in Georgia for the development of the shipment services;

Innovative shipment infrastructure is missing in the food shipment system and the level of use of digital technology is low.

It is necessary to

• Develop effective mechanisms for political support for food shipment, provide funding policies to ensure a unified chain of shipping and food markets;

• Harmonize rules and standards of managing food shipment with the regulations of other countries, especially with those of European countries;

• Digitalize shipment system, transport corridor and food markets in Georgia;

• Promote investments in the field of shipment and logistics in order to further improve the transport infrastructure;

• Develop transport and logistics centers based on geographical locations of the food markets;

• Improve the security policy of shipped goods;

• Integrate international and regional transport systems;

• Activate economic diplomacy, promote the transit potential of the country by holding respective forums, business meetings and exhibitions, etc.

References

1. United Nations Economic and Social Commission for Asia and the Pacific. (2015, July 9-10). Concept of Sustainable Transport: Planning and Designing for Sustainable and Inclusive Transportation Systems. National Capacity Building Workshop on Sustainable and Inclusive Transport Development Colombo, Sri Lanka 9-10 July 2015. UNESCAP Transport Division. https://www.unescap.org
2. United Nations Center for Regional Development. (n.d.). Environment - Environmentally Sustainable Transport (EST). https://www.uncrd.or.jp/index.php?menu=376
3. Martins V.W.B., Anholon R., Quelhas O.L.G. (2019). Sustainable Transportation Methods. In: W. Leal Filho (ed.) Encyclopedia of Sustainability in Higher Education. Springer, Cham. https://doi.org/10.1007/978-3-030-11352-0_192. https://link.springer.com/referenceworkentry/10.1007%2F978-3-030-11352-0_192
4. Erkomaishvili, G. (2019). Rational economic policy - Driving force from business to social entrepreneurship . Universal. http://dspace.nplg.gov.ge
5. Danelia I. (2019). Development perspectives of the transit opportunities of geoeconomic hub in Georgia. [Doctoral dissertation, Ivane Javakhishvili Tbilisi State University]. University Repository. https://www.tsu.ge/assets/media/files/48/disertaciebi4/Irakli_Danelia.pdf
6. U.S. agency for International Development. (2017, 22 December). Enhancing dialogue platform between public and private sectors in parliament. Research report: Transport and logistics field in Georgia: main challenges and development prospects. USAID Project "Governance for Development". https://pdf.usaid.gov/pdf_docs/PA00TH43.pdf

7. Campanhola, C., Pandey, Sh. (Eds.). (2019). Sustainable Food and Agriculture, FAO & Elsevier's Academic Press. https://www.elsevier.com/books/sustainable-food-and-agriculture/campanhola/978-0-12-812134-4
8. Sporleder, T. L., Boland, M. A. (2011). Exclusivity of Agrifood Supply Chains: Seven Fundamental Economic Characteristics. The International Food and Agribusiness Management Review, 14(5), 28-51. DOI: 10.17059/2013-4-18 https://www.researchgate.net/publication/227354964_Exclusivity_of_Agrifood_Supply_Chains_Seven_Fundamental_Economic_Characteristics
9. Diosey, R. L. M. (2020). Global Food Security in a Pandemic: The Case of the New Coronavirus (COVID-19): Review. World (1), 171–190.
10. Kharaishvili, E., Natsvlishvili, I. (2019). Markets for Georgian agri-food products and economic policy for export stimulation. Globalisation and Business, (8), 44-53 (in Georgian).
11. Kharaishvili, E. (2017), Directions for improving the supply chain in the agro-food sector of Georgia. In O. Garafonova (Ed.). Foresight management: formation and transformation adaptive business organizations. International collective monograph. (2, 6. pp. 159-169) Kherson: Publishing house Vyshemyrskyi V. S. https://www.tsu.ge/data/file_db/economist_faculty/shota.pdf
12. National Statistics Office of Georgia. (2020). Share of food expenditures in total consumer spending. Statistical Yearbook of Georgia: 2020. https://www.geostat.ge/media/35685/Yearbook_2020.pdf
13. Pinstrup-Andersen, P. (Ed.) (2015). Food price Policy in an Era of Market Instability: A Political Economy Analisis. Oxford University Press. https://library.oapen.org/bitstream/handle/20.500.12657/32080/617900.pdf?sequence=1&isAllowed=y
14. FAO Committee on World Food Security (2011, July). Price Volatility and Food Security: A Report by High Level Panel of Experts on Food Security and Nutrition. http://www.fao.org/3/mb737e/mb737e.pdf
15. Gelashvili, B. (2020, 19 June). Transportation and warehouse services become more expensive. Report New Agency. https://report.ge/economics/satransporto-da-sasatsqobo-momsakhureba-gadzvirda/ (in Georgian)
16. Organisation for Economic Co-operation and Development (OECD). (2020a April). OECD updates G20 summit on outlook for global economy. 27/03/2020 - Updated 15 April 2020. http://www.oecd.org/newsroom/oecd-updates-g20-summit-on-outlook-for-global-economy.htm
17. The World Bank. (2020a, May). World Bank Group: 100 Countries Get Support in Response to COVID-19 (Coronavirus), 19 May, 2020. IBRD-IDA/World Bank Groop. https://www.worldbank.org
18. The World Bank. (2020b). Annual Report 2020. Supporting Countries in Unprecedented Times. IBRD-IDA/World Bank Groop. https://www.worldbank.org/en/about/annual-report#anchor-annual
19. Natsvlishvili, I. (2020). Perspectives Of Sustainable Development And Sustainable Business Models In The Post-Covid Era. Globalization and Business, (10), 227-232. https://doi.org/10.35945/gb.2020.10.029 (in Georgian)
20. Rodrigue.J.P. (n.d.). The Logistics of Global Food Systems. The Geography of Transport Systems. The Spatial Organization of Transportation and Mobilit. https://transportgeography.org/contents/applications/logistics-global-food-systems/
21. Liu, S. (2017). Transport and agricultural productivity: A cross-national analysis. Research on Modern Higher Education. (2) Article 01012 https://www.researchgate.net/publication/314219295_Transport_and_agricultural_productivity_A_cross-national_analysis

22. Wajszczuk, K. (2016). The Role and Importance of Logistics in Agri-Food Supply Chains: An Overview of Empirical Findings. Logistics and Transport, (30), 47-56 https://www.researchgate.net/publication/305478060_The_Role_and_Importance_of_Logistics_in_Agri-Food_Supply_Chains_An_Overview_of_Empirical_Findings
23. Gray, R. S. (2020). Agriculture, Transportation and the COVID- 19 crisis. Special Issue: COVID-19 and the Canadian agriculture and food sectors: Thoughts from the pandemic onset, 68 (2), 239-243.
24. Aday, S., Aday, M. S. (2020). Impact of COVID-19 on the food supply chain. Food Quality and Safety, 4 (4), 167–180. https://doi.org/10.1093/fqsafe/fyaa024
25. Organization for Economic Development and Cooperation (OECD). (2020b June 2). Food Supply Chains and COVID-19: Impacts and Policy Lessons. OECD Policy Responses to Coronavirus (COVID-19). http://www.oecd.org/coronavirus/policy-responses/food-supply-chains-and-covid-19-impacts-and-policy-lessons-71b57aea/
26. Borsellino, V., Schimmenti E., Bilali H. El. (2020). Agri-Food Markets towards Sustainable Patterns. Sustainability. 12(60), 2193. https://doi.org/10.3390/su12062193
27. Soma, K., Bogaardt, M.-J., Poppe, K., Wolfert, S., Beers, G., Urdul, D., Pesce, M., Kirova, M., Thurston, C., Belles, C. M. (2019). Impacts of the digital economy on the food-chain and the CAP: Research for AGRI Committee. https://www.europarl.europa.eu/RegData/etudes/STUD/2019/629192/IPOL_STU(2019)629192_EN.pdf
28. Gebresenbet, G., Bosona, T. (2012) . Logistics and Supply Chains in Agriculture and Food., In Groznik, A., Xiong, Y. (Eds.). Pathways to Supply Chain Excellence. DOI: 10.5772/25907. https://www.intechopen.com/books/pathways-to-supply-chain-excellence/logistics-chains-in-food-and-agriculture-sector
29. Ackerley, N., Sertkaya, A., Lange, R. (2010). Food Transportation Safety: Characterizing Risks and Controls by Use of Expert Opinion. Food Protection Trends, 30(4), 212–222. http://www.foodprotection.org/files/food-protection-trends/Apr-10-Ackerley.pdf
30. Pal, A., Kant, K. (2017). A Food Transportation Framework for an Efficient and Worker-Friendly Fresh Food Physical Internet. Logistics 1 (10). doi:10.3390/logistics1020010 www.mdpi.com/journal/logistics
31. Dalkmann, H., Sakamoto, K. (2012). Urban Transport: Low Carbon Green Growth Roadmap for Asia and the Pacific. Policy recommendations for the development of eco-efficient infrastructure. https://www.unescap.org/sites/default/files/7.%20Urban-Transport.pdf
32. Berg, C. N., Deichmann, U., Liu, Y., Selod, H. (2017). Transport Policies and Development. The Journal of Development Studies, 53(5). https://www.tandfonline.com/doi/full/10.1080/00220388.2016.1199857
33. Martincus, C. V., Carballo, J., Cusolito, A. (2017). Roads, Exports and Employment: Evidence from a Developing Country. Journal of Development Economics, 125, 21-39 .
34. IPES Food-International Panel of Experts on Sustainable Food Systems. (2015, May). The New Science of Sustainable Food Systems: Overcoming Barriers to Food Systems Reform. Report 01. The Case for a New Science of Sustainable Food Systems. http://www.ipes-food.org
35. Ministry of Economy and Sustainable Development of Georgia. (2021). Economic outlook, Transport. http://www.economy.ge/?page=ecoreview&s=26
36. Ministry of Economy and Sustainable Development of Georgia. (2011, August 18). Order #1-1/1562 of the Minister of Economy and Sustainable Development of Georgia about adoption of technical reglament "rules of cargo shipment by autotransport". http://girca.org/uploads/____122.pdf

37. Metskhvariashvili, S. (2020, April 15). Shipment became more expensive-will it impact products' prices? Business Media Georgia. https://bm.ge/ka/article/gadazidvebi-gadzvirda---aisaxeba-tu-ara-es-produqciis-fasze-/54089
38. Government of Georgia. (2020, March 23). Resolution №185 of Georgia's government about adoption of state program on maintaining prices on basic food products. https://matsne.gov.ge/ka/document/view/4832948?publication=0
39. Dorosh, P., Wang, H. G., You, L., Schmidt, E. (2011). Road connectivity, population, and crop production in Sub-Saharan Africa. Agricultural Economics, 43(1), 89–103.
40. Bosker, M., Garretsen, H. (2012). Economic Geography and Economic Development in Sub-Saharan Africa. The World Bank Economic Review, 26(3), 443-484.
41. Ghani, E., Goswami, A. G., Kerr, W. R. (2016). Highway to Success: The Impact of the Golden Quadrilateral Project for the Location and Performance of Indian Manufacturing. The Economic journal, 126 (591), 317-357. https://doi.org/10.1111/ecoj.12207
42. Kvasha, S., Pankratova, L., Koval, V., & Tamošiūnienė, R. (2019). Illicit financial flows in export operations with agricultural products. Intelellectual Economics, 13(2), 195-209.
43. Kvach, Y., Piatka, N., & Koval, V. (2020). Management of sustainable entrepreneurship adaptation to tax changes in environmental investment. Baltic Journal of Economic Studies, 6(5), 96-105. https://doi.org/10.30525/2256-0742/2020-6-5-96-105
44. Derii Zh., Koval, V., & Sedikova I. (2018). The role of the agricultural sphere in the context of food security. Scientific Bulletin of Polissia, 4(16), 21–27. http://dx.doi.org/10.25140/2410-9576-2018-4(16)-21-27

Intermodal transportation as a quality improvement tool in tourism industry

Anzor Devadze[1] *, *Paata* Chaganava[2], *Irakli* Kordzaia[3], and *Lela* Devadze [2]

[1]Batumi Shota Rustaveli State University, Tourism Department, 35/32 Ninoshvili/Rustaveli str., 6010 Batumi, Georgia
[2]Batumi Shota Rustaveli State University, 35/32 Ninoshvili/Rustaveli str., 6010 Batumi, Georgia
[3]Batumi Shota Rustaveli State University, BSU Quality Assurance center, 35/32 Ninoshvili/Rustaveli str., 6010 Batumi, Georgia

Abstract. The international tourism, with its developed transport infrastructure, transforms the earlier closed community into the open one, where the communications between the representatives of different countries become the everyday routine. Georgia has strategic geopolitical location on the conjunction of Europe and Asia and holds all the abilities to develop different types of passenger transportation and it enhances the further development of international tourism in the country. The authors examine the condition of country's transport infrastructure and organization of passenger transportation via different types of transport, aiming to carry out the directions of improving the quality of tourist's transport service and raising the competitiveness of tourism industry in Georgia, the authors proposed Intermodal Transportation Service Center as a tool of improving the quality of tourism services.

1 Introduction

Georgia with its Geopolitical location and natural, historical and cultural resources has practical unlimited resources to develop international tourism. While analyzing the tourism development issues it is very important to define its interconnection wit transport industry, keeping in mind that the development of tourism and transport is the interconnected and interdependent process. Related to the above mentioned it is rather actual the identification of impact of transport infrastructure on development of international tourism in Georgia, as the role and importance of tourism is acknowledged as most important factor of international tourism development. It is natural and logic, because tourism is relatively new social-economic phenomena and mostly is the result of transport development [1-5]. It became very important after the significant qualitative and quantitative changes in the tourism flow volumes, dynamics and structures at national and international levels.

The aim of the given research is to perform complex analyze of development of international tourism and problems of organization of passenger tourist transportation via all types of transport in Georgia and based on it carry out the methodic and practical recommendations to raise the effectiveness of organization of international tourism in the

* Corresponding Author: devadze.anzor@bsu.edu.ge

country. To reach the goal of the research following objectives were set: analyze modern conditions of international tourism in Georgia; analyze modern conditions of organization of passenger transportation via different types of transport; identify problems in organization of passenger transportation; develop scientific-methodic recommendations for improving passenger transportation in Georgia aiming to raise the quality of transportation of international tourists via different types of passenger transport and competitiveness tourism industry of Georgia on the global market of tourism services.

2 Methodology

The theoretical and methodical base of the research are the publications of Georgian and foreign researchers in the sphere of economics, transport and tourism management, also the legal acts and normative-legal documents, regulating the legal and organizational issues of transport and tourism performance.

The informational base of the research are the materials of National Statistics Office of Georgia and its territorial bodies, Ministry of internal affairs and Ministry of Economy and Sustainable Development, also the information from Georgian National Tourism Administration and Tourism and Resorts Department of Autonomous Republic of Adjara. The methods of economic analysis, synthesis and analysis, and statistical methods were used during research. For the purposes of quantitative evaluations the official materials of National Statistics Office of Georgia were used to perform comparative analysis between analytical and statistical estimations. Practical importance of developed scientific-methodical and practical recommendations comprises the argumentation for systematic measures to improve the organization of performance of transport infrastructure of Georgia supporting the raising the effectiveness of development of international tourism in the country. Interesting publications on the topic, especially about the multimodal transportation and tourism are made by Efthymiou and Papatheodorou [6], Lohmann and Pierce [7], Yang, Li and Li [8], Chang and Shieh [9] and Darmawan and Chen [10].

3 Tourism and transport in Georgia

In the market economy the development of tourism business means the existence of recreational resources, suitable capital, technology and workforce. In contrast with other fields of economy, tourism resources of Georgia are much diversified and include natural and anthropogenic geo-systems, also the natural phenomena that have comfortable characteristics and consumer value for commercial activity to use for organization of leisure and recreation [1]. In the recent years (all the years except 2020, because of pandemic covid-19 that destructively influenced on international tourism, lowering its qualitative, quantitative and financial characteristics to the minimum. As a result, the article presents all the figures of international tourism development except 2020 year) Georgia faces sharp raising the number of international tourists and revenues from international tourism. in the year of 2019 the quantity of international tourists reached 9 357 964 persons (growth rate compared to the 2018 consisted of 7,8%). Also, the revenues from international tourism exceed 3 268,7 million US dollars in 2019 (Table 1).

Table 1. Dynamics of international tourist arrivals and revenues 2011-2019 [3].

Characteristic	**Years**						
	2013	**2014**	**2015**	**2016**	**2017**	**2018**	**2019**
International Arrivals, thousand.	5392,3	5515,6	5901,1	6350,8	7554,9	8 679,5	9359,0
Revenues, mln USD	1719,7	1787,1	1935,9	2097,4	2700,2	3222,0	3 268,7

The Table 2 below shows top-10 countries originate higher tourist arrivals in Georgia.

Table 2. Top 10 countries of international arrivals [2].

Country	**International arrivals in Georgia, Thousand**		**Growth, Thousand**	**Growth, %**
	2018	**2019**		
Azerbaijan	1,424,610	1,526,619	102,009	7.2%
Armenia	1,268,886	1,365,048	96,162	7.6%
Turkey	1,268,886	1,365,048	96,162	5.3%
Russia	1,404,757	1,471,558	66,801	7.6%
Ukraine	158,160	189,210	31,050	19.6%
Iran	280,735	137,223	-143,512	-51.1%
Israel	154,764	202,370	47,606	30.8%
Kazakhstan	51,489	77,856	26,367	51.2%
Poland	65,361	86,716	21,355	32.7%
Germany	61,422	85,251	23,829	38.8%

Positive tendency exists with the European union countries: for 2019 number of visits from European Union countries consisted 484 996, with the growth rate of 25,9% related to previous year. Germany was the first country in the list with the addition of the number of foreign tourists - +24,6 thousand and Polland - +21,4 thousand [3].

Quantity of accommodation objects, registered in the database of Georgian National Tourism Administration, consists of 2575 units, and the number of beds – 94438. The leader is the region of Adjara – 26519 (28,1%) beds, followed by Tbilisi – 23596 (25%) beds. There are several international brand hotels operating in Tbilisi: Radisson Blu Iveria Hotel, Tbilisi Marriott, Courtyard Marriott, Sheraton, Holiday Inn, Citadines Apart Hotel, Best Western Tbilisi, ercure, Hotels and Preference, The Biltmore Hotel, Ibis Styles Tbilisi Center, Best estern Tbilisi City Centre, Ramada Encore, Moxy by Marriott. The list enriched with Wyndham Grand и Ibis Tbilisi Stadium in 2019. International brand hotels are functioning in the regions also. In particular, in 2019 Radisson BLU Tsinandali, Park Hotel Tsinandali and Holidey Inn opened in Kakheti, Best Western Gudauri in Gudauri. According to the figures of STR global the rate of occupation for Georgia consisted 56,7% in 2019 [2]. According to the Ministry of Internal Affairs of Georgia, international tourists cross the state border of the country by different means of transport (Figure 1).

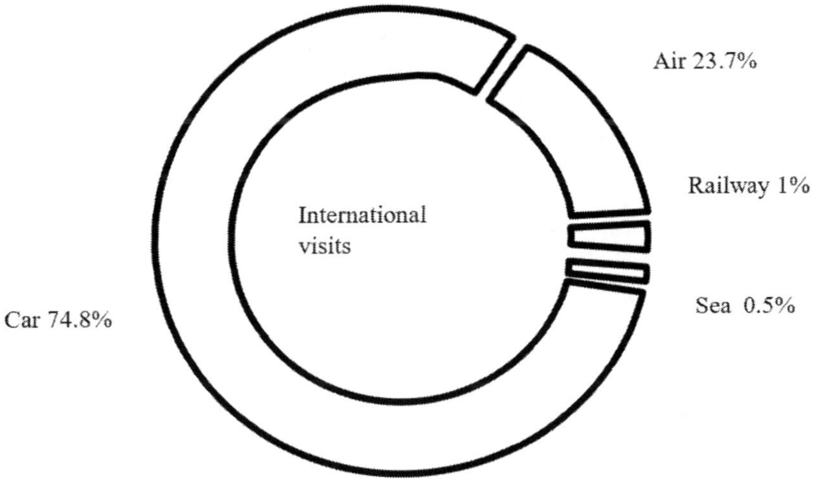

Fig. 1. Georgian border crossing by the different means of transport [4].

The most of crossings of international tourists assigned to the cars could be explained with the majority of travelers from neighboring countries: Armenia, Azerbaijan, Turkey, Iran etc., who prefers to travel with cars because of territorial closeness of Georgia and the accessibility of transport system. As for railway and sea transports' low rates, it is worth to mention that Georgia has the railway connection with Armenia and Azerbaijan only, also the sea passenger transportations take place from the sea ports of Batumi and Poti to the directions of Sochi (Russia) and Odessa (Ukraine), with the small amounts.

Georgia's flight market shows the considerable growth rate last years. In the pre-pandemic period there were 3 international and 2 internal airports working in Georgia, fully suitable to the standards of the International Civil Aviation Organization (ICAO). The international flight mostly consists of flights from and to Tbilisi airport. The management of Tbilisi and Batumi airports is performed by Turkish company TAV Airports Holding Co.

Air companies that worked in Georgia in Pre-pandemic period performed low-budgetary flights with relative prices: Wizz Air, Air Arabia, Pegasus, FlyDubai, Pobeda Airlines, Air Baltic, Buta Airways, Salam Air, Flynas, Ukraine International Airlines, Skyup Airlines.

Processing ability of Tbilisi, Kutaisi and Batumi airports consists 6,1 million, 600 thousand and 600 thousand passengers per year, accordingly. After the completing the expanding of airports in Kutaisi and Batumi (planned at 2021) their processing ability will rise to 2,5 million and 1,4 million passengers per year, accordingly.

In 2019 nine new international air companies enter the Georgian market, including French Air France, Armenian Aircompany Armenia, Maltese Air Malta, Irish Ryan Air, and Uzbek Uzbekistan Airways. Charter flights are performed by Thailand's Thai AirAsia X, Latvian GetJet Airlines and Turkish Onur Air. The table 3 contains the dynamics of international flights and passengers served for 2012-2017 from the Georgian airports.

Table 3.Quantity of flights and passengers served for 2012-2017 [3].

Years **Figures**	**2013**	**2014**	**2015**	**2016**	**2017**	**2018**	**2019**
Number of flights	13999	13576	13667	15318	20955	24606	24386
Passengers served, thsd.	1833,8	2008,5	2261,1	2840,5	4074,0	5033,3	5200,5

Transport is the key element of tourism product, in its part that the tourist consumes out of the tourist destination, on the way to it. While researching tourism, it is crucially important to define its interdependence with transport industry. The success on the tourist markets and adequate transport infrastructure are important conditions for development of tourist destination. By the other hand, demand on tourism is strong stimulus for accelerating the development of transport industry. Tourism depends on transport, its safety, speed and comfort; those are offered to the tourists during journey.

As it could be concluded from the analysis above, as tourism becomes massive phenomenon the range of the problems appears related to the transport services. One of the most important problems is the problem of imbalance of development of transport system of Georgia in total. It contains three elements.

Disproportion in the rates and scales of development of different types of transport. Best example for that – considerable low rates of sea transport development and high growth rates of automobile transport.

Insufficient development of existed transport infrastructure, that is shown in incomparability of the levels automobile roads to the level of automobile transport development and demand on automobile services, also in the existence of many problematic crossroads on the crossroads of different types of transport. Territorial inequality of development of transport infrastructure.

4 Intermodal transportation

Today the development of transport system becomes the necessary condition of implementation of innovation model of economic growth of Georgia and raising the service quality of passengers. Although the positive tendencies in the dynamics of several types of transport, the transport system in total doesn't respond to the existed demands and perspectives of country's development.

Each type of transport, performing passenger service, acts separately, according to its own interests, aiming to get maximum profit and do does not care about other types of transport. In sum the transport means are not used effective enough, the quality of passenger transportation services are low and their demand for transportation is not satisfied in sufficient manner.

If the role of transport infrastructure is not valued enough it affects negatively on the development of international tourism and quality of transport service of passengers and as a result, country could lose the image of attractive tourism country. To keep the level it is necessary to improve the existed legislation, perform the goal-oriented policy in the sphere of transport system.

Tourism and transport development is the interconnected and interdependent process. As it appears from the scientific literature the study of interconnections in the system "tourism-transport" the preference is given to the transport, because its role and meaning is acknowledged as an important factor of tourism development. It is natural and logic, as the tourism is relatively new social-economic phenomenon and in significant volume is the result of the transport development [5].

It is worth to mention that the majority of resorts in Georgia are in a significant distance from the major transport nodes of country, those are the cities of Tbilisi, Kutaisi and Batumi. , and the passengers voyaging to the resort zones face a number of problems when making the changes in the transport type to reach the final destination.

related to the above mention to improve the level of satisfaction of passengers, especially transit ones, in the transportation process there is a necessity of technical, technological, organizational and economic coordination of transportation process by different types of passenger transport, especially in key transport junctions of the country where the main mass of passengers, including tourists, are changing the transport types.

To meet the growing demands of tourists and passengers for transportation, it is desirable that the land transport agency, jointly with agencies of sea transport and civil aviation, under the management of Georgian ministry of economy and sustainable development to decide about creation joint web of service-center with the goal of elaborating and implementing the intermodal transport system on the main directions of passenger flows (Fig. 2).

Intermodal passenger transportation should be considered as the transportation of passengers, luggage and handbag from the point of origin to the final destination with the several types of transport by united transportation document, when the responsibility for total transportation process, including the connection in the conjunctions, takes particular operator or the third party operator (for example, tourist company).

Organization of intermodal passenger transportation represents the innovative and prospective directions of transport field development. This technology allows to combine the advantages of each type of transport and to make the transportation process most effective.

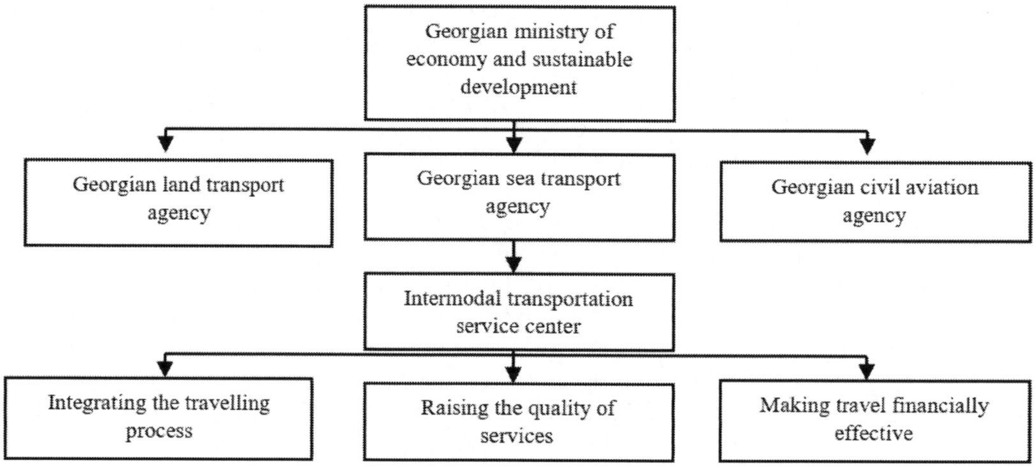

Fig. 2. Intermodal transportation service center.

In relation with the specifics of technology of organization of this type of passenger transportation, also taking into account the different requirements of passengers according to the types of transportation it is mostly preferable to use intermodal technology for the transportation of long haul passengers using such a types of transport as air, rail, car and sea transport. The key condition for such a technology is the presence of responsible person for successful change in conjunctions that means the solution of problems and avoiding of financial lost for passenger those could appear at the time of unsuccessful change, and the responsible person takes responsibility to solve the unforeseen situation and take the passenger to the final destination.

During the introducing the intermodal transportation it is desirable to draw the key principles to implement:

- High level of integration operators managing different types of transport;
- Information assurance of passengers during all the time of transportation “from door to door”;
- Unified system;
- Interaction of the transport types on the basis of united rules and requirements;
- Maximization of use of advantages of each transport type participant in intermodal transportation;
- United informational area;
- Value benefit for the passenger compared to the total value of separate transportation along the route.

Unification of above mentioned transport agencies into one web of service-center and establishing the suitable communications between them allows creating different tourist products according to the formulae “train + bus + . . .”; “train + bus + hotel”; “train + bus + excursion”; “train + bus + football event”; “airplane + bus + sanatorium” “airplane + bus + mountain resort” etc. advertisement of the created tourist products and their development could be performed by above mentioned service-center, tourist firms and agencies, in the close collaboration with sea, rail, car and air transport companies (Fig. 3).

The service-center could allow the improvement of the interaction of different transport companies that means improvement of quality of transportation of transit passengers on the basis of the elaboration of common transport schemas with agreed schedules of intercity or suburb buses with sea, rail or air transport at the transport junctions.

The specific attention should be paid to the some problematic issues, which appear during the service performed by tourist firms and transport organizations.

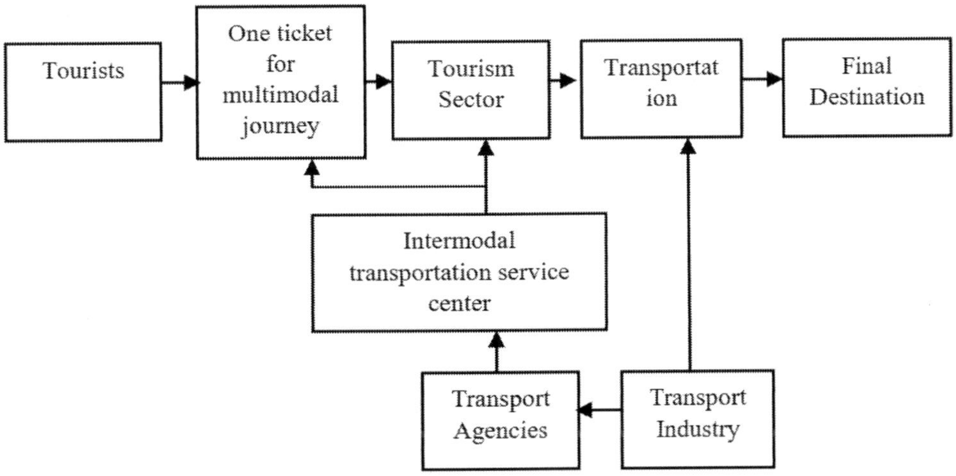

Fig. 3. Role of Intermodal transportation service center in the tourism services.

The most important is that the transport costs contain the 40-50% of total expenditures of tourist services.Such a ratio of prices blocks the tourist firm's ability to perform the elastic price policy during implementation of full pack of tourist service. In given conditions the non-price methods of competition are put on front and contain attracting consumers by improving quality of services offered alongside with additional services.

5 Conclusion

The problem of improving the effectiveness of transport infrastructure is one of the most important for the future development of country's international tourism. It should be considered on the comprehensive basis, taking into account the regional conditions of functioning of all transport types, reflecting the territorial and other specifics of their exploitation. Improvement of the quality of services for transit tourists should be ensured on the basis of elaboration of high-effective forms of transportation process organization by improving the interaction of different transport companies.

References

1. Devadze, A. (2012). Tourism in Georgia: existed situation, problems, perspectives. Publ.House Kabadoni.
2. Georgian National Tourism Administration. (2020). Account of 2019. Tbilisi.
3. Georgian National Tourism Administration. (2020). Operative Data of 2019. Tbilisi.
4. Ministry of internal affairs of Georgia. (2018). Operative Data of 2017. Tbilisi.
5. Meladze, M. (2008). The specifics of craft of tourist product in Georgia. Publ. House Kabadoni.
6. Efthymiou, M., Papatheodorou, A. (2015).Intermodal passenger transportation and destination competitiveness in Greece. Anatolia. 26:3, 459-471 https://doi.org/10.1080/13032917.2015.1012171
7. Lohmann, G., Pearce D.G. (2010). Conceptualizing and operationalizing nodal tourism functions. Journal of Transport Geography, 18, 266-275.
8. Yang Y., Li D., Li X. (2019). Public transport connectivity and intercity tourist flows. Journal of travel research, 58(1). 25-41.

9. Chang, P., Shieh, H. (2016). Seamless service strategies for passenger transportation in Taiwan. International Journal of Business and social science, 7, 136-148.
10. Darmawan, V.E.B., Chen, Y.W. (2020). Implementation of connection scan algorithm in tourism intermodal transportation journey planner: a case study. Jurnal system dan manajemen industry, 4, 129-136.

MATEC Web of Conferences **339**, 01003 (2021) https://doi.org/10.1051/matecconf/202133901003
ISTSML 2021

Risk management in maritime transport of goods through insurance

Milen Mitkov[1*], *Sergey* Radukanov[2], and *Mariana* Petrova[3]

[1]Air Force Academy „Georgy Benkovski", Dolna Mitropolia, Bulgaria
[2,3]St. Cyril and St. Methodius University of Veliko Tarnovo, Veliko Tarnovo, Bulgaria

Abstract. The present study, the issues related to the development and features of Cargo insurance are considered as a way to manage the risk in the maritime transport of goods and cargo. The state, tendencies and prospects for development of this type of insurance are outlined, as well as the gross domestic product of the country. The study period is 2010-2019, ie the years after Bulgaria's accession to the European Union. The results of the analysis confirm the strong link between maritime transport and insurance of transported goods and cargo, which leads to strong development in the insurance market as a whole. It was established that the rate of development of the gross domestic product of the country lags behind the rate of development of the transported goods and cargo by sea. On the other hand, the pace of development of freight insurance outpaces the pace of development of goods transported by sea and the rate of development of the country's gross domestic product.

1 Introduction

A lot of businesses are looking for innovative ways to improve efficiency and maintain competitive advantage in order to survive [1-3]. The economic purpose of maritime transport is to be transported from one place to another created by other mirror products, as the processes of relocation and consumption of its useful effect are photographed and compared over time. Unlike rail and road transport, maritime transport cannot be equipped anywhere in the country. It can be used only in areas that have suitable natural conditions - the presence of seas, large navigable rivers, conditions for digging artificial canals and others.

Many Bulgarian and foreign authors have worked and expressed their opinions competently on the problems of the economy as a whole and insurance. In this field we can distinguish part of the research of Pukala [4, 5], Mitkov [6], Bacho, Hkibko, Vnukova, Pola [7], Sushchenko [8], whose multi-layered analysis has enriched and expanded knowledge in this area. In this way solutions to important problems of economic and managerial practice are given from the standpoint of modern scientific achievements.

The problem of organizing efficient economics and innovation activity invoked great scientific interests of Ukrainian and Latvian researchers, such as Ponomarenko, Malyaretz,

* Corresponding author: m.petrova@ts.uni-vt.bg

Dorokhov, Matyushenko, Hlibko, Pasmor, Loktionova, Jarmusevica, Ilisko, Badjanova, Jukss, Komarova, Selivanova-Fyodorova, Ruza, Kaźmierczyk, Arbidane [9-12].

Compared to other modes of transport, water has the ability to perform mass international transport of goods, and during the transport there are many risks. Therefore, the study of the risks associated with the carriage of goods by sea is of particular importance. One way to manage these risks is through insurance of transported goods and cargo. Through it, the risk is transferred from the sea carrier to the insurance company. The role of insurance for the maritime carrier of goods is reduced to indemnification of those insured who suffer losses in the realization of risks included in the liability of the insurer. Insurance protection is therefore a support tool for entities that bear the risk associated with their activity [7].

The insurance provides insurance coverage covering partial damages or complete loss of cargo and related property interests during transport for loss or damage, which are a direct consequence of the occurrence of covered insurance risk. The object of the insurance are the goods transported by water, land and air. The goods are insured by pieces, parcels, connections or crates as indicated in the transport documents, and in the case of goods transported in bulk or in bulk - by weight units. The purpose of insurance is to transfer responsibility for the risk of loss of or damage to cargo from the buyer or seller of goods to the insurer.

The specialized nature of the Cargo insurance is an essential prerequisite for any increase or decrease in the economic activity in the country to reflect on the increase or decrease of the share of the insurance in the total volume of the concluded insurance contracts.

According to the definitions of "insured interest" adopted in world theory and practice, the property interest that the insurer has in relation to the insured object. Insurance is the optimal risk transfer instrument for an enterprise [4]. It is present when the insured has a direct material interest in the fate of the subject of the insurance. Depending on the conditions under which the specific delivery is agreed in the given transaction, the seller or the buyer, as well as the forwarders, banks, owners of transit warehouses, etc., may be interested in the insurance. Carriers are also interested in its existence, because although the insurance of the goods they transport does not release them from liability, it greatly reduces their risks from circumstances beyond their control. It should be emphasized that risk management is a very important tool in the operations of enterprises for monitoring and reducing the risk of operations [5].

The moment of the transfer of ownership of the goods from the seller to the buyer determines who is the bearer of the insured interest at the time, and hence who will be a party to the insurance contract. Given the change in ownership, the interest holder also changes. In the case of transport insurance, the bearer of the insured interest is always the person who bears the risks of loss of or damage to the transported cargo. Therefore, in commercial transactions, the moment for the occurrence or transfer of the insured interest between the participants in the transaction is directly related to the moment for the transfer of the risk for the goods from the seller to the buyer. On the other hand, the moment for passing the risk depends on the conditions of the specific delivery and the franking of the agreed price.

This study focuses on the development of cargo insurance in the context of risk management of the maritime carrier of goods and cargo, and the **subject** - on the economic, legal and technical possibilities, features and problems faced by maritime carriers in concluding of Cargo insurance.

The goal is to analyze and assess the state and trends in the development of cargo insurance in Bulgaria and to outline the trends in development.

In order to achieve a formulated goal, it is necessary to consider the following tasks:

– summarization and systematization of theoretical formulations for cargo insurance;
– analysis of the dynamics of the gross domestic product in Bulgaria for the period (2010-2019)
– analysis of the condition of the transported goods by sea for the period (2010-2019);
– analysis of the state of the insurance market in Bulgaria for the period (2010-2019);
– analysis of the condition of the cargo insurance for the period (2010-2019).

2 Materials and Methods

To achieve the main goal of the study, modern approaches and methods are used, such as: method of analysis and synthesis; inductive and deductive method, empirical approach, comparative method and logical description

All results refer to the respective sources of information of the National Statistical Institute of the Republic of Bulgaria, the Financial Supervision Commission, Eurostat or are based on the authors' own calculations. The main statistical data for the analyzes in the separate parts of the survey are taken from the official websites of: National Statistical Institute, Financial Supervision Commission and Eurostat. They cover the period until 31.12.2019 and do not take into account the changes that occurred after it. The data are presented in tabular form using MS Office Excel 2010.

All results refer to the relevant sources of information of the National Statistical Institute of the Republic of Bulgaria, the Financial Supervision Commission, Eurostat or are based on the authors' own calculations.

3 Results

3.1. Research and analysis in the development of the gross domestic product in Bulgaria for the period 2010-2019

One of the indicators for measuring economic activity is the gross domestic product. Along with labor productivity, it is one of the most important indicators determining the development of an economy. The most significant can be defined as the final result of the production activity of economic entities in the country. In today's rapidly changing socio-economic environment, the problem of sustainable development is becoming increasingly relevant as a concept for achieving environmentally friendly economic development of society. On the basis of the data obtained for the gross domestic product, the income of the population as well as its purchasing power can be estimated.

There are three main approaches to measuring gross domestic product: production, revenue and expenditure.

Gross domestic product, calculated by the production method, characterizes the final result of economic activity and is measured by the value added created in the production of goods and services by units resident in the economic territory of the country. Value added is the difference between the value of the final goods and the intermediate goods used to produce them. To the sum of the value added of all goods by producers in the national economy are added the net taxes (reduced by subsidies) and the value of the gross domestic product is obtained.

Under the production method, GDP at market prices is calculated as the sum of gross value added at basic prices total for the economy and adjustments, which include net taxes on products, non-deductible value added tax and import duties.

In the income approach, gross domestic product is presented as the sum of income received from resource providers, ie. by the owners of factors of production

Through the cost approach, gross domestic product is calculated on the basis of the monetary costs incurred by all producers of finished products.

The data in Table 1. and Fig.1 show that after 2010 the rate of development of the gross domestic product in Bulgaria is growing significantly. The most significant reason for this is the high growth of foreign direct investment during this period. In the period after 2010, the annual growth of gross domestic product per capita was on average about 1.8%. The slow but real growth of the economy after 2010 is largely due to the stable banking system, the currency board, as well as EU funds. The absorption of EU funds has a significant impact on the positive growth of the economy. By 2018, euro funds contribute to over 32% of total investments in the economy. Therefore, the funds from the Euro funds, which flow into the Bulgarian economy in the form of government expenditures, compensate to some extent for the outflow of foreign direct investment.

Table 1. Gross domestic product, growth rate for the period 2010-2019 (in %)

Years	GDP (billion BGN)	Compared to 2010	Compared previous year
2010	73,214	x	x
2011	78,221	6,8	1,6
2012	80,941	9,5	1,8
2013	81,357	10,9	0,5
2014	82,910	12,3	2,0
2015	86,792	17,8	3,3
2016	92,635	26,0	3.8
2017	102,300	39,7	3,6
2018	109,700	49,3	3,1
2019	118,670	61,6	3,4

Source: own calculations, based on National Statistical Institute of the Republic of Bulgaria

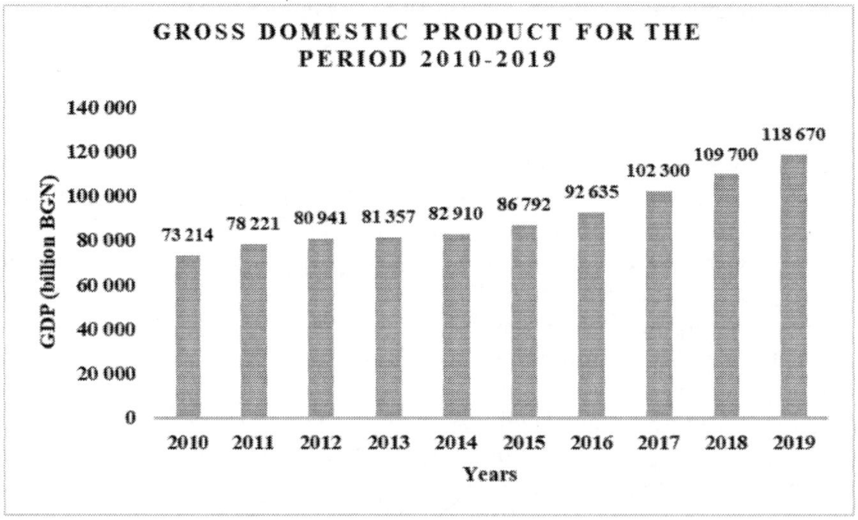

Fig.1. Gross domestic product for the period 2010-2019 (in BGN)

During the period 2010-2019 in Table 1, there is a continuous economic growth both compared to the base year and to each subsequent year. In 2019, compared to 2010, the gross domestic product increased by a little over BGN 45 million or by 62%.

There are several main objectives of economic policy, the priority of which depends on the specific state in which the economy is. These goals can be summarized in the following:

price stability (low inflation); high level of employment (or natural rate of unemployment); maintaining positive foreign trade balances and sustainable economic growth

3.2. Analysis of the condition of the transported goods by sea for the period for the period 2010-2019

The transported goods from the river and sea transport at the end of 2019 are 30 997 thousand tons, which is an increase of 11.2% compared to the previous 2018. From the data in Table 2 it is established that the export of goods and cargo has a significant share compared to the import of goods and cargo

Table 2. Goods and goods transported by sea in Bulgaria for period 2010-2019.

Years	Import unloaded /thousand tons/	Export loaded / thousand tons/	Total / thousand tons/	Growth rate %
2010	11847	11089	22946	x
2011	13036	12149	25185	9,8
2012	12722	13290	26012	3,3
2013	12503	16338	28841	10,9
2014	12299	14936	27235	-5,6
2015	12622	14544	27166	-0,3
2016	13115	15570	28685	5,6
2017	14184	16769	30953	7,9
2018	13046	14822	27868	-10,0
2019	14770	16227	30997	11,2

Source: own calculations, based on Eurostat

The data in Table 2 show that the rate of development of the transport of goods and cargo is ahead of the rate of development of the country's gross domestic product. The average rate of development of transported goods and cargo for the period from 2010 to 2019 is 7.6%.

The main characteristics of ports and inland infrastructure facilities, as well as the type of goods transported, are some of the elements that significantly influence the choice of maritime transport. The European Union is heavily dependent on seaports for trade with the rest of the world and within its internal market, accounting for 74% of trade (imports and exports) with the rest of the world.

Ports ensure the territorial connectivity of the countries of the European Union by servicing regional and local maritime traffic to connect with peripheral and island areas. They are the hubs from which multimodal logistics flows can be organized across the trans-European network, using short sea shipping links, railways and inland waterways to minimize the strain on road infrastructure and energy consumption.

3.3. Analysis of the state of the insurance market in Bulgaria for the period 2010-2019

A country's economy is a dynamic system that develops depending on the socio-political relations that dominate it. Creating everything necessary for the existence of society, it is one of the foundations for its development

In the process of economic development, the insurance activity also occupies its important place. The well-developed insurance sector is an integral part of any modern economic system, often it is part of stable economic growth, part of the high employment rate of the working population. In the highly integrated economic space of the European Union, the contribution of the insurance sector to the economies of the Member States is crucial.

The provided insurance protection against the consequences of adverse events enables the companies and the citizens to receive monetary compensation and to cope with the occurred difficulties. In this sense, insurance helps to achieve stability in the economies of individual countries. The main measure of the condition and development of the insurance market is the gross premium income. The economic essence of premium income is manifested in the fact that it forms part of national income. It is generated by the insured in order to guarantee their interests from the harmful effects of adverse events

Table 3. Premium income of insurance companies in Bulgaria for period 2010-2019.

Years	Non - live insurance	Live insurance	Total premium income	Growth rate	Share of Non-life insurance	Share of Life insurance
	Million BGN	Million BGN	Million BGN	%	%	%
2010	1 374,7	248,6	1 623,4	x	84,7	x
2011	1 362,0	251,7	1 613,7	-0,8	84,4	15,6
2012	1 336,0	268,0	1 604,1	-0,6	83,3	16,7
2013	1 423,4	305,9	1 729,4	7,7	82,3	17,7
2014	1 434,3	340,6	1 774,9	2,6	80,8	19,2
2015	1 573,0	391,2	1 964,3	9,9	80,1	19,9
2016	1 622,2	428,3	2 050,6	4,21	79,1	21,9
2017	1 748,7	428,6	2 177,4	6,2	80,3	19,7
2018	2 047,1	430,1	2 477,2	11,4	82,6	17,4
2019	2 383,3	477,3	2 860,6	11,6	83,3	17,7

Source: own calculations, based on Financial Supervision Commission

The data in Table 3 show that in 2019 the gross premium income of all insurance companies in the country reached nearly 2.9 billion. There is also a steady trend towards an increase in premium income for the entire period of the survey. The rate of development of premium income for the period 2010-2019 ranges from - 3.50% to 9.90%. In the years up to 2008 the rate of development ranges from 16.62% to 28.16%. The crisis in the period after 2008 also affected the insurance market.

The data in the table show that in the period from 2010 to 2012 there is a negative rate of development, which ranges from -0.60% to -3.50%. A good trend can be noted the positive pace of development of premium income after 2013, as at the end of 2018 and 2019. an increase of over 10% is observed. For the entire period of the study from 2010. until 2019 the rate of development of the premium income is 76.2%. The growth of premium income is mainly due to the liability under the Civil Liability Insurance, Credit and Guarantee Insurance

During the period 2010-2019. the tendency of permanent increase of the importance of life insurance for the development of the insurance market and the economy is confirmed, after which in recent years this importance decreases. The outpacing pace of development of life insurance compared to non-life insurance led to an increase in the share of life insurance in gross premium income. If the relative share of life insurance in relation to premium income was before Bulgaria's accession to the European Union within 10% or a ratio of 9:1 in favor of non-life insurance, then in 2017 the share is already about 20% or the ratio is 8:2. However, this is still far from the usual European practice, where life insurance is predominant.

3.4. Analysis of the state of cargo insurance for the period 2010-2019

The following presentation presents the development of Cargo insurance over a relatively long period, covering the period after the global crisis of 2007.

Table 4. Premium income on cargo insurance in Bulgaria for period 2010-2019.

Years	Non - live insurance Million BGN	Growth rate %	Cargo insurance Million BGN	Growth rate %
2010	1 374,7	x	13,7	x
2011	1 362,0	-0,9	15,2	10,9
2012	1 336,0	-1,9	14,6	-3,6
2013	1 423,4	6,5	15,8	8,4
2014	1 434,3	0,1	15,6	-1,2
2015	1 573,0	9,7	17,2	9,7
2016	1 622,2	3,1	21,2	23,3
2017	1 748,7	7,8	18,2	-14,1
2018	2 047,1	11,7	17,9	-1,7
2019	2 383,3	11,7	18,9	5,6

Source: own calculations, based on Financial Supervision Commission

The data in Table 4 show that in 2019 the gross premium income from cargo insurance exceeds BGN 18 million, reaching BGN 18.9 million. The following development trend is established: in the years from the initial research period the income increases, then gradually decreases, alternating years of positive and negative rate of development in 6 consecutive years. The pace of development of premium income for the period 20010-2019 ranges from - 14.1% to 23.6%.

The data in the table show that at the end of the study period in 2019 the revenue increased by over 5% compared to the previous 2018. The significant increase in trade and the favorable monetary policy pursued by the leading industrialized countries maintain high levels of trade. Thus, for the period 2010-2019 we observe a certain decline, which is due more to the choice of the owners of the transported goods to rely on another insurance contract - "Liability of the carrier", despite the limited coverage it provides. Due to its specific nature, Cargo insurance affects the interests of a relatively small circle of economic business units and entities. This inevitably leads to a decline, and sometimes to a lack of a solid and sustainable marketing policy by the licensed companies in order to promote it.

It should be noted that the decrease in the amount of collected premiums is due not only to the low levels of trade and the smaller number of concluded contracts, but also to the decrease in the prices of transported raw materials, which reflects on the insurance values, on the insurance amounts under contracts, and hence on the required premiums.

Table 5. Market share of cargo insurance in Bulgaria for period 2010-2019.

Years	Cargo insurance (Million BGN)	Market share (%)
2010	13,7	0,99
2011	15,2	1,11
2012	14,6	1,09
2013	15,8	1,11
2014	15,6	1,09
2015	17,2	1,09
2016	21,2	1,38
2017	18,2	1,21
2018	17,9	0,87
2019	18,9	0,79

Source: own calculations, based on Financial Supervision *Commission* www.fsc.bg

The data in Table 5 show that for the period 2010-2019 there is a steady trend towards maintaining the market share of cargo insurance within 0.8 - 1.10%. It is no secret that until recently, cargo insurance, especially for domestic deliveries, did not receive much attention. For several years, the trend has been to pay more attention to these insurances. When we have land transport, traders often save on cargo insurance, relying on the liability of the carrier, which sometimes leads to losses.

From the above it can be concluded that the development of sales of insurance "Freight during transport" is not directly dependent only on trends in the Bulgarian insurance market. The increase in sales and premium income under this type of insurance also largely depends on international trade and the tendency of exporters and importers of goods to transfer the risk to insurers.

4 Discussion

A country's economy is a dynamic system that develops depending on the socio-political relations that dominate it. In recent years, Bulgaria is facing a significant demographic crisis, which is characterized by declining and aging population. The country's gross domestic product is constantly growing, which is a prerequisite for sustainable economic growth. On the other hand, a stable and sustainable economy is a prerequisite for increasing foreign direct investment, which will lead to even greater development and turn the country as a regional factor in Central and Eastern Europe. In recent years, sustainable development has been established in the transport of goods and cargo by sea to and from the country. The results show that the rate of development of transported goods and cargo by sea significantly the rate of development of the country's gross domestic product.

The insurance market is an integral part of any modern economic system, and often it is part of stable economic growth. Based on the research, it was established that Cargo insurance has a huge potential for development in the coming years. Growing economic activity in the country and the related transport of goods and cargo by sea is expected to significantly increase the share of insurance of "Cargo during transport" or the so-called "Cargo" insurance, in the total volume of insurance contracts. The reason for this is that this type of insurance is directly related to the economic situation in the country.

5 Conclusions

Based on the study, it is established that Cargo insurance has great potential for development in the coming years. The growing economic activity in the country and the implementation of this transportation of goods and cargo by sea is expected to increase the share of insurance of "Cargo during transportation" or the so-called "Cargo" insurance, in the next volume of excluded insurance contracts. The reason for this is that this type of insurance is directly supported by the economic situation in the country. The policy cannot be sold in the country where there is no developed business, because it is used by sea carriers. The gradual stabilization of economic activity in the external environment has had a positive effect on exports of goods and services by sea.

A country's economy is a dynamic system that develops depending on the socio-political relations that dominate it. In recent years, Bulgaria is facing a significant demographic crisis, which is characterized by declining and aging population. There are several main objectives of economic policy, the priority of which depends on the specific state in which the economy is. These goals can be summarized in the following: price stability (low inflation); high level of employment (or natural rate of unemployment); maintaining positive foreign trade balances and sustainable economic growth.

References

1. Yankovyi, O., Goncharov, Y., Koval, V., & Lositska, T. (2019). Optimization of the capital-labor ratio on the basis of production functions in the economic model of production. Naukovyi Visnyk Natsionalnoho Hirnychoho Universytetu, 4, 134-140.
2. Mushkudiani, Z., Gechbaia, B., Gigauri, I., & Gulua, E. (2020). Global, economic and technological trends in human resource management development. Access Journal, 1(1), 53-60. https://doi.org/10.46656/access.2020.1.1(4) (2020).
3. Baklanova, O., Petrova, M., & Koval, V. (2020). Institutional transmission in economic development, Ikonomicheski Izsledvania (Economic Studies), 29(1), 68-91.
4. Pukala, R. (2016). Use of neural networks in risk assessment and optimization of insurance cover in innovative enterprises. E&M, 8 (3). doi.org/10.1515/emj-2016-0023
5. Pukala, R., & Petrova, M. (2019). Application of the AHP Method to Select an Optimal Source of Financing Innovation in the Mining Sector, IVth International Innovative Mining Symposium, E3S 105, 04034, https://doi.org/10.1051/e3sconf/201910504034.
6. Mitkov, M. (2018). Status and trends in the development of macroeconomic indicators "Insurance penetration" and "Insurance density" in Bulgaria (1997–2016). SEA, 10(1), 34-44.
7. Bacho, R., Pukala, R., Hlibko, S., Vnukova, N., & Pola, P. (2019). Information Management: The Key Driver of the Economic System's Development. Marketing and Management of Innovations, 3. p. 297-307. http://doi.org/10.21272/mmi.2019.3-23
8. Sushchenko, O.A. (2015). Global shifts as aspects of globalization impact on foreign economic activity management. Contribution to Intern.Economy, 1,41-48.
9. Matyushenko, I., Hlibko, S., Petrova, M. M., Pasmor, M. S., & Loktionova, M. (2020). Assessment of the development of foreign trade in high-tech production of Ukraine under the association with the EU. Business, Management and Education, 18(1), 157-182.
10. Jarmuševiča, V., Iliško, D., Badjanova, J., Jukšs, V., & Petrova, M. (2019). SMART governance of implementing the strategy of corporate societal responsibility for a sustainable regional development. Proceedings of the Int.Multidisciplinary Scientific GeoConference Surveying Geology and Mining Ecology Management, SGEM 19(5.3), pp. 645-652,
11. Ponomarenko, V., Malyaretz, L., Dorokhov, O. (2011). Genesis and methodological foundations of multidimensional data analysis in the social and economic system. Montenegrin Journal of Economics, 7 (2), 111-120.
12. Komarova, V., Selivanova-Fyodorova, N., Ruza, O., Kaźmierczyk, J., Arbidane, I. (2021. Modern trends in the economic differences between countries and within them: comparison of the world and the European Union, Entrepreneurship and Sustainability Issues, 8(3), 110-121. https://doi.org/10.9770/jesi.2020.8.3(6).)

Actual problems for logistics management and strategies of supply chain in Georgia

Leila Tsetskhladze[1*], *Nanuli* Makharadze[1], *Irma* Chkhaidze[1], *Nato* Jabnidze[1], and *Nazi* Baratashvili[1]

[1]Batumi Shota Rustaveli State University, Department of Business Administration, Management and Marketing, Ninoshvili/Rustaveli str. 35/32, Batumi, Georgia

Abstract. In social-economic development strategy, logistics is considered as one of the priority direction to increase the competitiveness of the country. In order to adapt the flexibility of the supply chain, the companies should properly work and carry out such logistics strategy, which allows them to identify the influence of the expected changes and make organizational or functional amendments to ensure the reduction in logistics costs. The development of logistics system contributes the integration of Georgia in global economy space and strengthens its competitive positions on international logistics market. In the work, it`s been studied the potential of increasing the functional load of Georgian ports and the potential of establishing them as logistics centers, it`s been reviewed logistics role, as a vital sector in terms of global trade and as a continuous and uninterrupted delivery of goods during a crisis period of Covid-19 pandemic. Strategies for supply chain has been studied and evaluated by sectoral mark according to expertise, economic-statistical and analytical methods and it`s been defined the role of Georgia in global logistics network and the actuality of involvement in it.

1 Introduction

The ongoing processes of globalization around the world and aspiration of integration with European Union creates the opportunities for active involvement of Georgia in international trade. In the mentioned process, transport and logistics sector must perform a crucial role of the country to become a part of global supply chain. This has a positive effect on the development of entrepreneurial and export activities, on increasing the accessibility of the market, education, employment and additional investment [1].

The future key elements of production and sales system and distribution network will be innovations, digital technologies, transport and logistics and for commercial organizations, online platforms, blockchain decisions and others. It`s essential the digitization of information exchange and interaction for continuity of maritime transport operations under the pandemic conditions. It ensured the continuity of transport operations and trade procedures. The rapid introduction of technological decisions ensures the continuity of entrepreneurial activities and state procedures.

* Corresponding author: Cecxladze.leila@bsu.edu.ge

The significant improvement of transport and logistics sector in Georgia has been conditioned by the economic and infrastructural reforms, carried out in the recent years in the direction of simplification of customs procedure and liberalization of services in key sector of economy.

As a result of it, the transit and logistics potential has been increased, what should have to create the conditions with global markets to improve the international connections. But it should be noted that, Georgia still faces the challenges in connection with the transformation as regional and logistics hub, which includes low quality of service, outdated logistics infrastructure and equipment, ineffective management of supply chain, lack of qualified staff and others [1].

While working on the work, the attention has been paid on the problems of supply chain strategy and logistics management in Georgia. Expert-specialists, employed in the field of logistics, evaluated the types of service via vehicles considering the relevant criteria and sectoral mark. The most ponderable (weighted) criteria was evaluated and revealed between the pre-selected criteria by using the expert method. According to economic-statistical, comparison, analysis and synthesis methods of the study it was identified the actual problems for management of Georgian logistics companies and it was determined the ways, which should promote the involvement of the research field in the global logistics network.

2 Results

The globalization of the economy increases its attitude to the supply mechanisms, transportation and logistics. The beginning of Covid-19 pandemic has confirmed the critical importance of transport and logistics systems not only on state level but international level too. According to the data of "Allied Market Research", logistics global market increases with 3,5 % per year, its value in 2019 year was 4,53 billion dollars, it should have to achieve 12,3 billion dollars by 2020 year and according to the prognosis, the mentioned value will reach to 16,82 billion dollars by 2027 year. Herewith, the average annual growth rate will be 17,9 % by 2020-2027 years [2].

Particular attention should be paid to the optimization of the supply chain management, by which the following problems are solved: reducing the whole cycle of planning; expanding its area due to the reliable and timely delivered information; the optimization of product types, expenses and suppliers, selection of the required contractors, support the relationships with them; minimization of production costs between the contractors via information exchange and products flows. Timely connections between the participants of the supply chain gives an opportunity to be revealed in advance so called "narrow places" directly in the production process, to reduce the warehousing costs according to the demand based on the optimization of production volume. This task responds to the concept of Just-in-Time delivery. To improve the quality of logistics service via flexible and timely delivery process. According to the data of the largest analytical companies (AMR Research, Forrester Research), the following advantages are reviewed by using SCM: profit growth from 5-to-15 %; reduction of order processing time and cost from 20-to-40%; reduction of warehouse stocks from 20-to-40 %; reduction of the time of placing on the market from 15-to-30%; reduction of purchase costs from 5-to-15%; reduction of production costs from 5-to-15%. Despite this, Operations and finance leaders have good reason to be fearful about the continued challenges supply chains will face in 2021 [3].

Instability of logistics system in Georgia causes high logistics costs, what is accompanied by lack of coordination in terms of order processing, stock management, warehousing and distributing. The logistics performance index (LPI) of the world bank indicates to the need of development-perfection of logistics system in Georgia, where according to the data of last 2018 year, Georgia had 119th place between 160 countries. The rating of 5-point scale is done in accordance with six main indicators: efficiency of customs procedures and border management; quality of transport infrastructure: ability to produce competitive international transportation; ability and quality of logistics services; possibility of cargo tracking; timely delivery to the place of destination according to the cargo schedule [3]. Deterioration was observed in traceability and tracking rates. The mentioned is connected to the lack of integration of information and communication technologies and electronic data exchange systems and low level of logistics processes.

Industry 4.0 plays an important role in logistics and supply chains management. The transition to digital manufacturing and online shopping forces us to take a fresh look at logistics, as a management instrument of value chain and determine the focus of expected changes in logistics. The analysis of the results of Covid-19 crisis should stimulate international dialogue, the goal of which will be the simplification of conducting the transactions between the borders [4]. As the practice, introduced in successful logistics systems, shows us one of the perspective direction for Digital SCM/Logistics is using the blockchain technology. Blockchain is multifunctional and multi-level information technology, the purpose of which is the reliable accounting of various assets and transactions. Potentially, this technology includes all fields of economic activity: finances and economy, transactions in tangible and intangible assets, state operations and accounting in the companies, management of supply chain and logistics and others. [14].

The use of this technology will profoundly change the operations of the company in the supply chain, because the clients will have the direct information about goods and services of company-contractors. Investors will be able to finance the companies without financial markets as well; business-models will be able to be changed and improved radically; the cost structure will become transparent and the execution of the contracts will become unhindered [1]. Blockchain gives us a new decentralized paradigm in integration and collaboration of supply chain, including in finances, information, products and technologies. Blockchain is able to increase the efficiency in automatically processing data and reducing the number of intermediaries, in automatic management [33].

Accordingly, Blockchain-technology integrated the supply chain and helps the business in cost reduction and improvement of collaboration via exchanging the information. It has enormous potential to stimulate the digitization of the field supply chain and to establish the common standards of collaboration. IBM and Maersk worked out the supply Blockchain called TradeLens, which is a platform, where there`s shown the sample of international supply chain. During the process, all users share the relevant current status, so each of them confirms the latest information [22].

Recently, the share of maritime shipment has been increased in the world cargo turnover. As table 1 shows us, dynamics of shipping by maritime transport is increasing. In 1994 year compares with 1989 year, more than 13 % shipping was made, in 2004 year compared with 1999 year – more than 18 % and in 2019 year compared with 2014 year – more than 12 % cargo shipping was made. If we compare 1979-2019 years – cargo turnover of marit8ime transport was increased with 189%, what is caused on the one hand by the growth of global turnover and on the other hand, with the advantage of using the maritime transport (Table 1). The main transport systems of Georgia are airports of international and local importance, seaports /terminals and rail transport. According to the data of ministry of economy and sustainable development, total amount of the cargo transported by the transport sectors in Georgia in 2020 year consisted – 42,9 million tons, this indicator has been increased with 0,6 million tons (42,3 million tons) compared with 2019 year and with 1,8 million tons (41,1 million tons) – compared with 2018 year. to total number of cargo, transported in transport sectors is characterized with the growth trends under the conditions of Covid-19 compared with the previous years, what is stipulated with uninterrupted functioning of transport sector (Fig. 1).

Table 1. dynamics of the loaded products on maritime transport according to the cargo type (million tons) [12].

Year	Total goods	Crude oil	Other tanker trade	Dry cargo
1974	3304	1497	335	1472
1979	3828	1725	318	1785
1984	3364	1079	399	1886
1989	3940	1260	468	2212
1994	4485	1498	509	2478
1999	5683	1553	532	3598
2004	6758	1770	546	4442
2009	7818	1710	931	5177
2014	9816	1712	1122	6983
2019	11076	1860	1308	7907

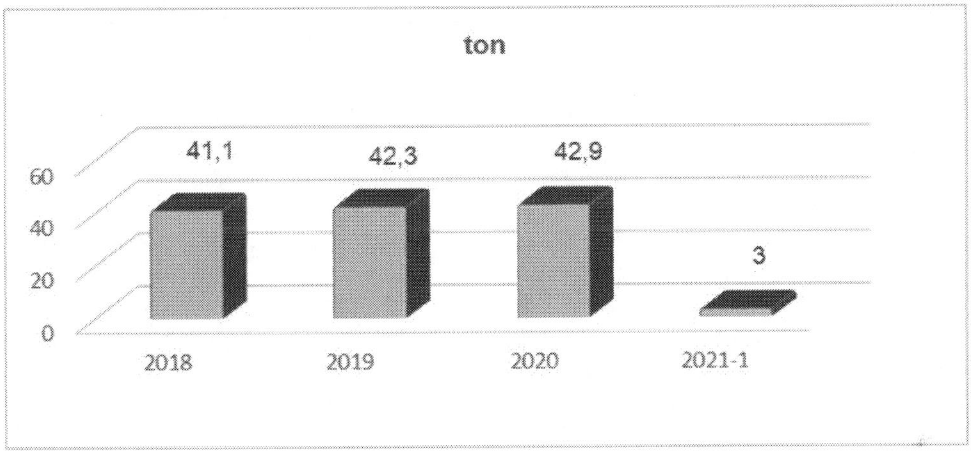

Fig 1. The (total) number of cargo, transported by transport sectors in Georgia.

There are four seaports open for international navigation in Georgia. All of them are owned or managed by private companies. All types of cargo (dry, container, liquid) are processed in Poti and Batumi ports. LPG, crude oil and petroleum can be processed in Kulevi oil terminal and crude oil and petroleum is processed in the port of Supsa.

Table 2. The cargo, processed in seaports of Georgia in 2019 year and statistics of throughput mastering [18].

Cargo and capacity (million tons)	Batumi		Poti		Kulevi		Supsa	
	processed cargo	mastering %	processed cargo	mastering %	processed cargo	mastering %	processed cargo	mastering %
Dry cargo handling	0,7	35%	3,4	85%	-	-	-	-
Ferry crossings	0,8	100%	0,3	29%	-	-	-	
Oil reloading	0,9	6%	0,6	56%	1,6	26%	3,7	46,3%
Containers, TEU	116,08	58%	531,7	97%	-	-	-	-

As table 2 shows us, the potential of maritime transport of Georgia is high. But its mastering indicators are not satisfactory. Due to the fact that a large part of the cargo flow in Georgian ports is transit, the reduction of economic activity in neighboring countries has a negative effect on dry and liquid cargo, also the redirection of cargo to other ports /corridors, operational restrictions due to configuration, problems related to depths (silting up) (Poti, Kulevi), what limits to use large tonnage vessels. Due to the above, seaport operators have to control the depth of the entrance canals and make deepening works, what is related to the significant costs [20]. The combination of high tariffs, trade imbalances and waiting for a long time at wharfs is a hindering factor for attracting the additional cargo flows to ports.

It should be noted that the role of maritime transport as a key sector has been increased around the whole world under pandemic conditions, which has a great importance in terms of continuous and uninterrupted delivery of vital goods and global trade during the crisis periods. Pandemic has become a kind of test stone not only for globalization, but also for global solidarity and cooperation. It`s necessary the agreed actions for data standardization, for monitoring of ports activity and for developing the defense mechanisms against cyber security [23]. The annual throughput of Georgian railway is 28 million tons. During 2013-2018 years, the number of the freight, transported by railway was decreased from 18.2 million tons to 10.0 million tons, but by 2019 year, it was increased with 9 % and throughput was provided with 39 %. It should be noted growing trend of container transportation of the rail, which is maintained despite of epidemic situation by 2020 year.

Strategic geopolitical location of Georgia stipulates the interest of neighboring countries as well as the interest of Europe and East Asian countries to use Georgia as a transit region. Trans-Caspian International Transport Route - "Middle Corridor" provides the freight transportation from China to Europe and backwards via Kazakhstan, Azerbaijan and Georgia. Southwest corridor has a great potential for Georgia, because it connects the countries of the Persian Gulf and India with the European Union countries through Georgia. The first railway container transportation of goods from China via "Middle Corridor" was carried out in 2015 year. It should be noted that, by passing the "Middle Corridor", there has been a sharp increase in the number of container transportations from China to the direction of Georgia and Europe, in particular, in 2019 year it was 2,774 TEU, what is more than 107.3% compared with 2018 year. Herewith, it should also be noted the fact that, transportation was also carried out in the direction of Gzech Republic, namely, the first container block train was passed from China, C. Syan and successfully arrived in C. Prague passing through Georgia and Turkey, in particular Baku-Tbilisi-Kars railway. There should be noted that despite world challenges and delays, existing in transport and trade sector, caused by spreading Covid -19 pandemic, the growing trend has been observed this year too. in particular, in 2020 year, in the period of January-July, 1,785 TEU was transported with the abovementioned route, what is more than 81,6 % (983 TEU) compared with the same year of 2019 year. [20].

According to statistical data of Civil Aviation Agency, the volume of the freight, transported by airline companies, acting in Georgia [10] totally consisted of 34,065 tons in 2016 year, what is increased with 18,799 tons (totally 15,266 tons) compared with 2010 year. Turnover was reduced with 2,341 and 8,554 tons according to the years or it consisted 31,42 tons in 2017 year and 25,511 tons – in 2018 year. The volume of the freight, transported via airways was 24 911.336 tons in 2019 year, what is less than 543.49 tons, accordingly less than 2.14 % than the number of the cargo, transported in the same period of 2018 year [10]. The number of the freight, transported via air ways was 19 266,220 tons in 2020 year, accordingly turnover was reduced with 22 % [11] (Fig.2). The reason for this is on the one hand the pandemic-related restrictions and on the other hand the high prices of cargo transportation vi air transport.

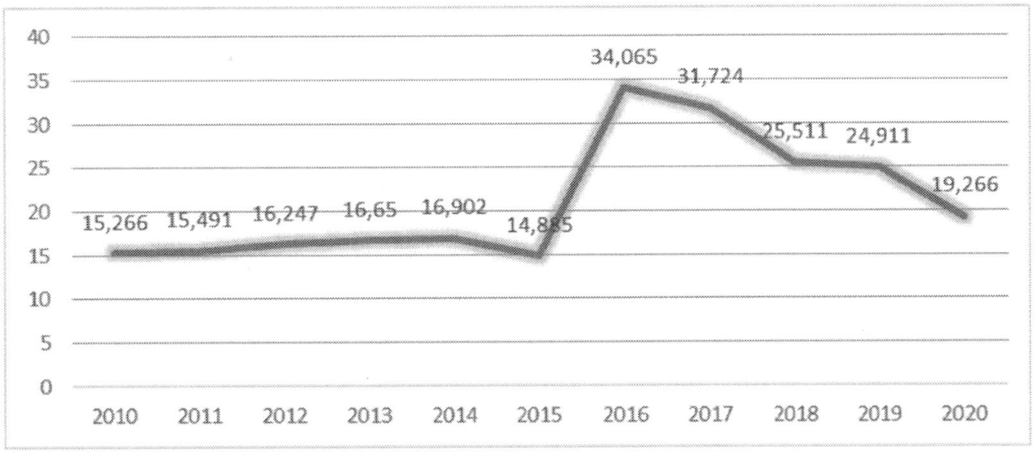

Fig. 2. Volume (tons) of the cargo, transported via air transport in Georgia

International freight transport has been characterized with stable dynamics. In 2019 year, 10,7 million tons freight was transported, what is higher with 16,8 % than the previous year indicator is and it totally consists of 25,2 % of freight transportations.

Table 3. The number of freight, transported via freight vehicles (thousand tons).

Mode	**2014**	**2015**	**2016**	**2017**	**2018**	**2019**	**2019/6**	**2020/6**
Export of Georgia	692.0	910.6	968.9	1430.8	1253.7	1513.0	723.9	619.5
Growth compared with the previous year	-	31.6%	6.4%	47.7%	-12.4%	20.7%	-	-14.4%
Import of Georgia	2975.8	2952.1	3299.0	3705.0	3894.7	4376.6	1871.7	1972.9
Growth compared with the previous year	-	-0.8%	11.8%	12.3%	5.1%	12.4%	-	5.4%
Transit of Georgia	2638.3	2708.4	2886.1	3628.5	3992.5	4783.2	2209.5	2509.2
Growth compared with the previous year	-	2.7%	6.6%	25.7%	10.0%	19.8%	-	13.6%
Total	**6306.1**	**6571.1**	**7154.0**	**8764.3**	**9141.0**	**10672.7**	**4805.1**	**5101.6**
Growth compared with the previous year	-	4.2%	8.9%	22.5%	4.3%	16.8%	-	6.2%

Thus, Georgia is a link between Europe and Asia and it has a chance to become transport and logistics hub and in order to achieve this, the following goals are set: the development of reliable, effective and quick logistics services, human capital, transport and logistics sectors, secure connections and corridor. This is stipulated with the market requirements and with the need to develop private sector services, with no alternativeness of their support.

For the purpose of searching the information about the problems, existing in logistics and supply chain, we conducted expert research, where participated the entities, employed in this sector. The conducted expert research gives an opportunity to evaluate the problem quickly, the basis of which is high qualification of experience of the professionals. We considered it appropriate to evaluate 10 experts, who were interviewed through the direct way within the marketing research.

Most of the experts prefer road transportation from the supply channels, sea, air and mixed shipments are less required. The selected supply channel is reliable and safe for six experts, but in most cases he has no other alternative. It should be noted that outscoring is used by only one carrier out of many. Most of them don`t use it or they have and exclusive partner.

Depending on the type of procurement, most companies use online platforms, some of them addresses to the networks of logistics companies, or the system of contractual agreements and personal contacts. According to the most carriers` opinion, the competition is great but in most cases, they`ve no competitors and they are only ones on the market.

The following was revealed from the problem of supply chain: reduced freight turnover, imposition of new regulations at customs points, insolvent clients, quoted permit, delayed consent for transport permit on oversized and special cargoes, queues at customs checkpoints, prolongation of freight sending and delivery period, reduction of passengers flow (what reduced the income with 90%), duration of time and damage of the cargo. The latest Covid-19 regulations around the world, the change of the exchange rate reduced the freight turnover and the number of insolvent customers was increased. Accordingly the orders were reduced with 10 % and freight turnover – with 20 %, also there was reduced the realization process and passengers movement. Types of services via vehicles was evaluated with five-point system by the competent specialists/experts, employed in logistics sector according to the criteria, shown in the table 4.

Table 4. The evaluation of criteria of service types via vehicles and the relevant weighting coefficients.

	Criteria (a_i)	**experts (j)**										**Sum**	Medium	**weightingc oefficients**
		1	2	3	4	5	6	7	8	9	10			
1	reliability	5	5	5	5	5	4	4	5	5	5	48	4,8	0,1599
2	terms of delivery	3	5	4	4	3	4	4	5	4	4	40	4	0,1336
3	frequency of cargo transportation	5	5	4	4	4	4	3	3	5	5	42	4,2	0,1398
4	transportation cost	5	3	3	3	5	3	5	3	4	4	38	3,8	0,1273
5	ability of freight transportation and delivery	5	5	5	5	3	3	3	3	5	5	42	4,2	0,1388
6	accessibility	5	5	5	5	5	2	5	5	5	5	47	4,7	0,1551
7	safety	0	5	5	5	5	4	5	5	5	5	44	4,4	0,1455
	sum	28	33	31	31	30	24	29	29	33	33			1,000

Based on the obtained results, for the purpose of calculating the criterion weight (24), the ponderable (weight) criteria was revealed.

$$ai=\frac{\sum(p_{ij}:p_{cj})}{m} \quad (1)$$

In which ai – is a moment of the parameter, i – is a number of the parameter; j – is a number of the expert; m – quantity of experts in the board; P_{ij} – the ball grade of the i parameter by j expert; P_{cj} - total of the ball grades of the parameters by the j experts.

As the obtained results show, the reliability weight criterion consisted of a_1=0,1599; The weight index of the delivery time for the next criterion is equal to 0,1336 (a_2=0,1336); the index weight of criterion of frequency of the third cargo transportation is equal to 0,1398 (a_3=0,1398); Accordingly the weight of the cost of the fourth transportation is 0,1273 (a_4=0,1273); Accordingly the possibility of freight transportation and delivery, the weighting coefficient is 0,1388 (a_5=0,1388); the weight of the access is 0,1551-ს (a_6=0,1551) and safe criteria weight is equal to 0,1455 (a_7=0,1455). According to the obtained results, it`s clear that the most weighted criteria is reliability, accessibility and safety, what is particularly important and actual under the conditions of Covid-19.

3 Conclusions

Conducted studies and experts` evaluation gave us an opportunity to form the advisory opinion regarding the development of logistics and transport sector:

- the reduction of infrastructural deficiencies, which hinders the international trade turnover;
- facilitate of intensive introduction of international practice of logistics and supply chain;
- rapid mastering and digitalizing of the technologies, herewith technologies increasingly reach in supply chains and its distribution networks, including in logistics. Making technological decisions and knowledge of the latest achievements will become an essential condition and not the option. Pandemic showed us that pioneers are able to withstand the fluctuations faster in technologies mastery. For example, e-commerce and online platforms, decisions made on the basis of blockchain.
- risk assessment and their management. The risks assessment should be integrated with global interconnections and interaction.
- staff training and retraining. Logistics is high-technological sector, which is suffered with a deep shortage of the experienced staff not only in Georgia but around the world as well.
- attract the investments and introduce multimodal transport infrastructure.
- agreed work between different kinds of transports, increase the productivity of the operations based on the growth of the qualification and efficiency of human capital.
- development of all components of transport network in time and in parallel from sender to receiver in order not to degrade the results of system operation and reducing the efficiency of system operation, what make the sector uncompetitive.

References

1. Ministry of Economy and Sustainable Development of Georgia (2017). National strategy of transport and logistics of Georgia for 2021-2030 years. http://www.economy.ge
2. Allied Market Research (2020). Supply chain analytics Market statistics-2027. https://www.alliedmarketresearch.com/supply-chain-analytics-market
3. Supplychain247 (2021). Three Fears Over Supply Chainsin 2021. https://www.supplychain247.com/paper/three_fears_looming_over_supply_chains_in_2021/elementum
4. Kutelia, B. (2020). Critical Infrastructure Protection: transport and logistics, Economic research center, Tbilisi,. supply chain, Trends Food Sci. Technol., 91, 640–652. (2019).
5. Dengen, D., & Lalvan, Ch. (2016). Global management and supply chain management. JohnWiley&SonsLtd.
6. United Nations, UNCTAD, (2019). Digital economy report, Value creation and capture: implications for developing countries.
7. Chikovani, E. (2020). The main directions for improvement of freight transportations in Georgia, Tbilisi.

8. Irannezhad, E. (2019). Is blockchain a solution for logistics and freight transportation problems? World Conference on Transport Research – WCTR, Mumbai.
9. World Bank Document (2014). Georgia: opportunities to achieve the prosperity. Economic memorandum of the country.
10. GCAA (2018). Annual Report 2018. https://gcaa.ge/wp-content/uploads/2021/03/GCAA-annual-report-2018.pdf
11. GCAA (2020). Annual Report 2020. https://gcaa.ge
12. UNCTAD (2020). Handbook of statistics. https://unctad.org/webflyer/handbook-statistics-2020
13. Margalitadze, I. & Gegeshidze, E. (2016). Concepts for transport policy and maritime affairs, Tbilisi.
14. Chang, J., Katehakis, M., Melamed, B., & Shi, J. (2018). Blockchain Design for Supply Chain Managemen, TSSRN Electronic Journal. https://www.researchgate.net/publication/330034181
15. Duan, J., Zhang, C., Gong, Y., Brown, S., & Li, Z. (2020). A content-analysis based literature review in blockchain adoption within food supply chain. International journal of environmental research and public health, 17(5), 1784.
16. Arvis, J. F., Ojala, L., Wiederer, C., Shepherd, B., Raj, A., Dairabayeva, K., & Kiiski, T. (2018). Connecting to compete 2018: trade logistics in the global economy. World Bank.
17. Swan, M. (2017). Anticipating the economic benefits of blockchain. Technology innovation management review, 7(10), 6-13.
18. Ministry of Economy and Sustainable Development of Georgia. http://www.economy.ge/?page=ecoreview&s=26
19. Wang, M., Wu, Y., Chen, B., & Evans, M. (2020). Blockchain and supply chain management: a new paradigm for supply chain integration and collaboration. Operations and Supply Chain Management: An International Journal, 14(1), 111-122.
20. Jabbari, A., & Kaminsky, P. (2018). Blockchain and supply chain management. Department of Industrial Engineering and Operations Research University of California, Berkeley.
21. Rožman, N., Corn, M., Požrl, T., & Diaci, J. (2019). Distributed logistics platform based on Blockchain and IoT. Procedia CIRP, 81, 826-831.
22. Hackius, N., & Petersen, M. (2017). Blockchain in logistics and supply chain: trick or treat?. In Digitalization in Supply Chain Management and Logistics: Smart and Digital Solutions for an Industry 4.0 Environment. Proceedings of the Hamburg International Conference of Logistics (HICL), Vol. 23 (pp. 3-18). Berlin: epubli GmbH.
23. UNCTAD (2020). Reviev of Maritime Transport. https://unctad.org/system/files/official-document/rmt2020_en.pdf
24. Fatkutdinov, R. A. (2000). Strategic Marketing, Moscow.
25. Kamilaris, A., Fonts, A., & Prenafeta-Boldύ, F. X. (2019). The rise of blockchain technology in agriculture and food supply chains. Trends in Food Science & Technology, 91, 640-652.
26. Charaia, V., Anguridze, O., Gulbany, Sh., Kurdghelia, L., Lashkhi, M., Shatakishvili, D. (2020). The China Factor in Georgian Economy, Tbilisi.
27. Kupriyanovsky, V., Sinyagov, S., Klimov, A., Petrov, A., & Namiot, D. (2017). Digital supply chains and blockchain-based technologies in a shared economy. International Journal of Open Information Technologies, 5(8), 80-95.
28. UNCTAD (2021). Technology and Innovation Report. https://unctad.org/webflyer/technology-and-innovation-report-2021

29. UNCTAD (2020). Trade and Development,. https://unctad.org/topic/macroeconomics/trade-development-report
30. DHL Trend Research. (2018). Blockchain in Logistics, Perspectives on the upcoming impact of blockchain technology and use cases for the logistics industry.
31. The World Bank (2020). Connecting to Compete 2018 Trade Logistics in the Global Economy. https://openknowledge.worldbank.org/bitstream/handle/10986/29971/LPI2018.pdf
32. Agustina, Y., Winarno, A., Pratikto, H., Narmaditya, B. S., & Filianti, F. (2020). A Creative Economy Development Strategy: The Case of Trenggalek Creative Network for Trenggalek Regency, Indonesia. The Journal of Asian Finance, Economics, and Business, 7(12), 1111-1122.
33. Deloitte (2017). When two chains combine Supply chain meets blockchain, Deloitte. https://www2.deloitte.com/content/dam/Deloitte/us/Documents/strategy/us-cons-supply-chain-meets-blockchain.pdf

Problems of the formation of marketing communications in the field of management of logistics systems

Gela Mamuladze[1*], *Davit* Katamadze[1], *Guliko* Katamadze[1], *Giorgi* Mamuladze[1], and *Ivane* Bakuradze[1]

[1]Batumi Shota Rustaveli State University, 32/35 Rustaveli/Ninoshvili st. Batumi, 6010, Georgia

Abstract. The article considers the problems in the field of logistic system management and analyzes the role of marketing communications in solving these problems. Based on the research, it was found that marketing communications can improve the connections established between the individual links of the logistics chain, which determines the effectiveness of the entire economic system and the degree of integration. Relevant research has not been conducted yet in the country, therefore, effective utilization of the concluded recommendations will help to raise the level of management of transport hubs in Georgia. The results of the study can be used by companies working in the field of international maritime transport, air, rail and road. It is assumed that if European and Asian producers increase the shipment of their products through our country from Europe to Asia or vice versa, the effective use of Georgia's transit capacity (function) will be guaranteed. The methodological basis of the article is bibliographic studies of Georgian and foreign scientists in the field of logistics systems management and marketing, data collection, observation, survey (in-depth interview). The method of data collection is mainly used in local logistics companies, where the situation with the use of marketing communications was analysed.

1 Introduction

Marketing communications in logistics companies can be established in informational, direct, reversible, and other forms. In order to build relationships with freight owners (or users of transportation services) logistics companies often use direct marketing, public relations and advertising. In logistics systems, the connection between transportation and warehouse companies is much closer than their connection to other (external) economic systems. Establishing a flexible marketing communication mechanism between the supplier and the customer is extremely important for the effective management of the logistics system. Effective marketing communications should ensure an increased level of integration of transport channels. The low efficiency of the logistics systems in Georgia is explained by the low level of marketing communications between its transport channels (rail, road, sea and air) [5].

* Corresponding author: mamuladze.gela@bsu.edu.ge

Improving the management of the logistics system requires the establishment of a close relationship between its constituent structural elements, which should lead to a high level of integration and organization. Representatives of commerce, logistics and marketing are involved in a unified process of logistics system management. The exchange of manufactured products, their placement in wholesale, retail networks and sales is the common goal of their activities. The involvement of stakeholders in the process of distribution-exchange of goods is profit-oriented, which is characterized by elements of commerce, marketing and logistics [8].

For the research, we examined the activities of 15 B2B products and service companies operating in Georgia, from 15th of January to 10th of April, 2021. Also, taking into account the interests of the research, the views of representatives of various logistics companies were analyzed.

Management of the logistics system includes the control of the movement and placement of material flows, the integration of the individual logistics process into unified system to minimize the delivery time of the goods to the right place at a low cost. However, the challenge is that unified system requires the mutual economic interests of the individual entities. Under such conditions, it will be possible to achieve a result-oriented approach to the process of movement of goods.

A comprehensive logistics approach to transportation management supports optimization of the cargo flows, reducing the overall cost of freight. In addition to reducing costs in the delivery and shipping processes, it is necessary to reduce costs in the packaging of goods. The retail chain needs to establish a direct partnership between all participants in the logistics process.

One of the ways of proper management of the logistics system represents periodic, continuous updating of marketing communications between its components (warehouses and transport nodes). This goal can be achieved by providing technical, technological, economical and mathematical solutions to existing problems in the field of marketing communications, which ensures the proper management of the logistics system.

To improve the management of logistics systems and control the movement of goods, managers need to simultaneously plan the movement of material flows as well as organize operations and inventories [10].

Planning of material flows includes the control of the logistics company budget, operational and calendar planning of transportation processes, warehouse network, control of information systems and production of final products. Inventory management includes demand forecasting, control and management of inventories of finished products, semi-finished products, raw materials.

2 The results of the research and discussion

If the development strategies of the individual logistics companies are well aligned and successfully heading toward a common goal, this integration will lead to a more thorough result than the uncoordinated use of marketing communications. If the marketing approaches of the individual parties involved in the logistics chain are not unified in handling common challenges, the potential customers might be lost. The lack of such coordination in many Georgian logistics companies creates the compatibility problem in the marketing approach [2]. For example, transport companies may emphasize the efficient and in time delivery of goods, while warehouse managers may underline safe storage of goods. How will these different approaches in marketing communications create a sustainable and solid logistics system and strengthen its position in the market?

An uncoordinated approach in marketing communications is destructive for a logistics company and negatively affects its image, competitiveness. Especially if the marketing messages carry contradictory information.

For example, if a logistics company offers improvement of safe handling of the goods, on the one hand, and reduced freight rate will lead to losing their trust within the customers.

One of the indications of marketing communication issues among the logistics companies is a dissimilarity in the tariff policy. The transportation tariff, as an economic category, includes the cost of cargo delivery, which should provide coverage of operating costs and generate a certain profit. One of the factors influencing the choice between available delivery options is the cost of delivery. If there is an opportunity for alternative freight routes between countries, it is possible to revise transport and tariff policies in terms of reducing delivery costs.

It can be achieved as a result of the development of competition [9]. For example, in the case of alternative transit routes between Iran and Azerbaijan-Armenia is developed in the area of freight flows between the regions of Europe and Asia, Georgia faces serious competition. Due to the territorial conflict between Azerbaijan and Armenia over the Karabakh and the negative attitude of the US and Western European countries with Iran because of the construction of nuclear power plant, Georgia has had the opportunity of full utilization of its transit function. However, considering the fact that relation with Iran is improving, this opportunity has slightly decreased.

Most likely, the current situation will reduce the volume of transited cargo through Georgia. This is based on an analysis of data provided by the Ministry of Economy and Sustainable Development of Georgia on Georgia's transport channels in 2016-2020 [9, 11-13].

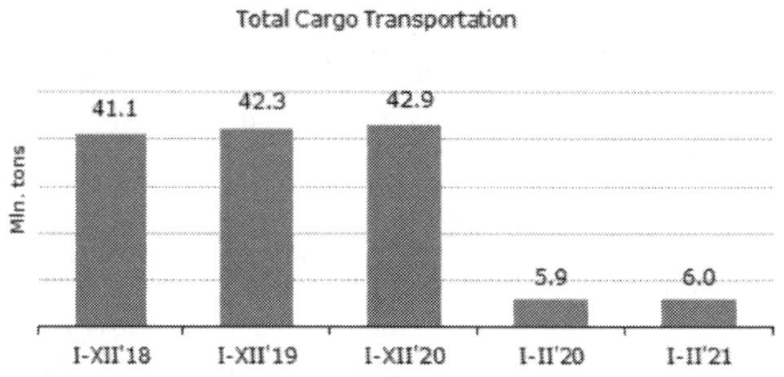

Fig 1. Total Cargo Transportation [13].

As shown in Figure 1, the amount of cargo transported by Georgia's transportation sectors from 2016-2020 was steady at 41.1 million to 42.9 million tons. Despite the spread of the COVID-19 pandemic around the world in 2020 (including Georgia), the amount of cargo transported through Georgia's has increased by 0.6 million tons, compared to the same figure in 2019.

In order to harmonize the marketing communications within the transport hubs of Georgie, it is necessary to unify and synchronize tariff policies. It should be noted that processing cargo flows through the country for certain time is supported by long-term contracts which are usually agreed between European and Asian suppliers and logistics companies operating in Georgia. In addition, freight tariffs can be maintained at a relatively high level by creating demand for additional shipment services [4]. In fact, the Georgia Department of Transportation had significantly reduced the protection and accompanying tax in February of 1994, which increased demand for services. As a result, the total amount of transported goods was increased.

Different types of transport channels have completely different tariff systems. In particular, common, special, preferential and local tariffs are used to determine the cost of cargo transported by rail in Georgia. The cost of the transportation of bulk cargo is determined by the common tariff.

Special tariffs are derived from general tariffs with a special surcharge or discount, which is applied to a particular freight flow and affects the development of Georgia's industry so that it can help regulate the inter-industry balance of the economy. The special tariff can be used to maintain the same transportation time by railroad at different seasons of the year. Reduced tariffs on damaged cars and containers serve the same purpose. The preferential tariff is used for special purpose freight - transportation by rail. Local communication tariffs are set by the heads of railroad departments. These tariffs include the amount of freight and rates of the various sets operating on that railroad. In addition to the cost of delivery, the railroad charges the shipper and consignee for additional transportation services. This charge is called a bundle and is charged for the following railroad operations: storage, weighing and inspection of cargo, delivery and cleaning of wagons, disinfecting them, forwarding, loading and unloading.

The amount of the freight tariff by rail is determined by the following criteria: 1. Type of cargo (wheat, oil, furniture, car); 2. Volume of cargo flow; 3. Type of wagon (universal, specialized or insulated wagon, container, tanker, platform); 4. Wagon capacity (60 tons, 25 tons, 10 tons); 5. Delivery speed (freight, oversized and passenger); 6. Delivery distance; 7. Owner of the vehicle (railroad, consignee, shipper) and others.

Transportation of goods by road is taxed at the following rates: freight rate; Price per ton/hour, distance rate, tariff agreement. Trucking rates are affected by the range, weight, and volume of the cargo. The capacity of the vehicle must be considered in determining the rate for transported goods based on following indicators: total distance, time of usage of the vehicle, type of vehicle, area in which the shipment occurs, and other factors. When transporting goods by trucks, each tariff does not take into account the entire set of factors, but some of them that apply in the case of a particular shipment. For example, to calculate the cost of delivery by freight rate close attention should be paid to the weight and class of cargo, which characterize the capacity of the vehicle, as well as its use and total milage. In all cases, the tariff is affected by the area of shipment. It explains the difference in the cost of freight delivery by region, which is adjusted by a correction factor.

As regards the freight rate for the goods transported by sea, it is determined by the factors, such as: area of trading, types of cargo, the amount of the cargo, types of the ship used, etc. For liner shipping that transports the goods at regular routes on fixed schedules, the freight rate is preliminary stated. On the other hand, for the tramp services that is a ship that has no fixed routing or itinerary or schedule and is available at short notice (or fixture) to load any cargo from any port to any port, the cost of transportation is set at the freight rate.

In logistics hubs, where cargo is transported by several types of transport mode, the supplier has to deal with and pay multiple rates separately [7]. Therefore, by using proper marketing communication with representatives of different parts of the logistics system, the supplier will be able to negotiate over the key aspects of the transportation, such as, to improve the safety of freight flow, to schedule, choose place and time, reduce delivery costs. At the same time, the cargo owner can choose the logistics company according to the quality, efficiency, safety and delivery rate, and change the types of transport from one point to another to make the best combination. In this respect, those transit countries and logistics companies which can simultaneously offer several types of transport means and routes, have a higher level of competitive advantage, as this allows customers to choose more alternative freight transport that simplifies their task.

3 Discussions

At the present moment, in the context of globalization and internationalization of the world economy, improvement of the management of logistics systems requires the development of a high level of marketing communications and the use of marketing intermediaries in the transport channels. The advanced marketing communications between individual elements of logistics systems will help accelerate the turnover of invested capital. Effective marketing communications play an important role in the exchange and distribution of information, financial and material flows of products. Intermediaries by using effective marketing communication tools will be able to support the synchronization, harmonization and unification of cargo flow as well as identification of challenges in logistics hubs.

In order to identify the existing problems in the field of marketing communications in logistics systems, a qualitative research has been conducted - through surveys and in-depth (telephone) interviews. Within the framework of the research, 15 logistics companies operating in Georgia were selected. The results of a qualitative survey showed that logistics companies are failing to properly utilize the potential of marketing communications.

From this point of view, it was interesting the opinion of logistics companies concerning the development of the logistics industry in Georgia. According to the survey, 66.7% of respondents believe that the sector is "partially developed", while 22,2% and 11.1% of respondents claim as „developed" and "not developed", respectively.

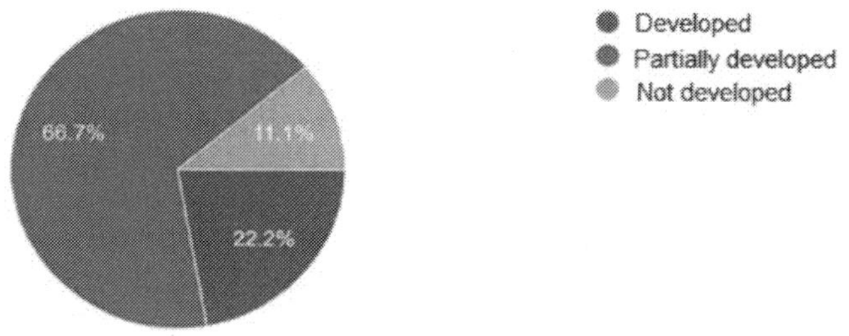

Fig. 2. Respondents' views towards the development of the logistics market in Georgia.

In order to study the research topic in detail, it was necessary to determine what factors hold back the development of the logistics market. According to the survey, the majority of respondents (66.7%) consider the problem to be the "lack of appropriate personnel". it is followed by the "lack of appropriate infrastructure", the third - "legal framework", the fourth - "lack of adequate financial resources", the fifth - "low level of transport development," and the sixth - "low level of warehousing.

The research revealed the necessity to identify factors constraining the development of logistics companies. The results of the study showed that the most negative impact on the development of logistics companies has a "Covid pandemic" (66,7%). In terms of impact, it is followed by "changes in the Georgian national currency GEL exchange rate", the third place by "high transport tariffs", and the fourth one by "tax tariffs".

Besides, the identification of factors dependent on freight rates, is under a spotlight. According to 77.8% of the respondents, freight rates depend on a distance. In terms of influence on freight rates, the second place is held by "cargo type", the third place by "cargo quantity" and the fourth one by "shipping time".

It was also interesting to determine the factors that define the logistics companies' competitive advantage. The majority of respondents (66.7%) consider such factor as "time". As regards to the rest factors, such as "opportunities", "productivity", "resources" and "result", the respondents perceive as same. Respondents give relatively less importance to "skills" and "costs".

Another circumstance was revealed as the need to identify the measures taken by companies to improve their reputation. All respondents clearly stated that "reliability" and "adequacy of price/quality" were the most important in this regard. They are followed by "integrity", then "access to information", fifth and sixth places by "charity" and "loyalty". None of the respondents considered usage of "sponsorship event" to enhance their reputation.

It was also interesting to note the attitude of logistics companies towards the use of marketing communication forms. In this case, as in the previous question, all respondents prefer "phone" and "e-mail" approach. "Website" (44,4%) holds the second place, the third place - "social networks", fourth and fifth places - "chat" and "Skype", respectively.

The research revealed another key element: it was interesting to know to what extent logistics companies use marketing communication tools in their activities? (Figure 3). 44.4% of respondents said they "do not use" at all, 44.4% use these tools partially, and 11.1% use them systematically.

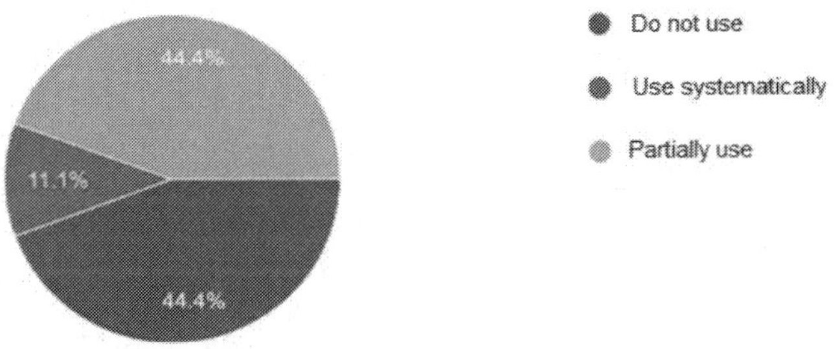

Fig. 3. Respondents' attitudes towards the use of marketing communication tools.

Source: The authors' research.

The research also showed interesting reasons why logistics companies refuse to use marketing communications, the survey results showed that 57.1% of respondents reject to use marketing communications due to their inefficient application. Others refuse because the use of these tools requires significant financial expenses.

In this regard, it was interesting to find out what marketing communication tools are most often used by the logistics companies we surveyed (Figure 4). The survey showed that 42.9% of respondents use "direct sales," 42.9% use public relations (PR), and 14.3% use advertising. None of them use "sales incentives".

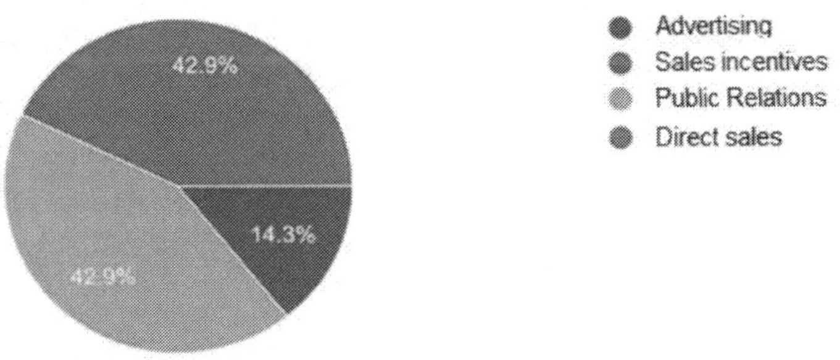

Fig. 4. Marketing communication tools used by logistics companies.

The survey presents an interesting overview of how logistics companies operating in Georgia have used advertising as an effective tool for marketing communications (Figure 5).

The survey found out that 71.4% of respondents most often used email as an advertising tool, 14.2% used catalog distribution, and 14.2% used none of the above. Surprisingly, not a single company uses Internet and media advertising tools.

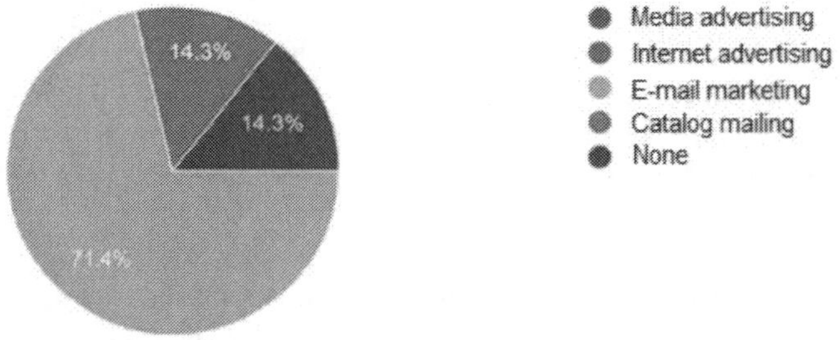

Fig. 5. The use of advertising by logistics companies as a tool for information distribution.

To fully analyze the current challenges, it was necessary to determine to what extent logistics companies in Georgia used PR as a marketing tool (Figure 6). The survey showed that Annual reports (20%), Seminars (20%), Presentation (20%) and charitable activities (20%), were evenly used, while the rest 20% do not use any of the above tools. Surprisingly, not a single company uses the press as a public relations (PR) tool. These and similar factors indicate that marketing communication tools are not properly implemented in the management of Georgia's logistics system.

Fig. 6. The use of PR-mailing by logistics companies.

Besides, it was interesting to specify how logistics companies use direct marketing in their activities as a tool to promote their brand. The survey showed that 75% of respondents use email marketing and 25% use telephone marketing. As for catalog sales, none of the respondents use them as a direct marketing tool (see Chart 7).

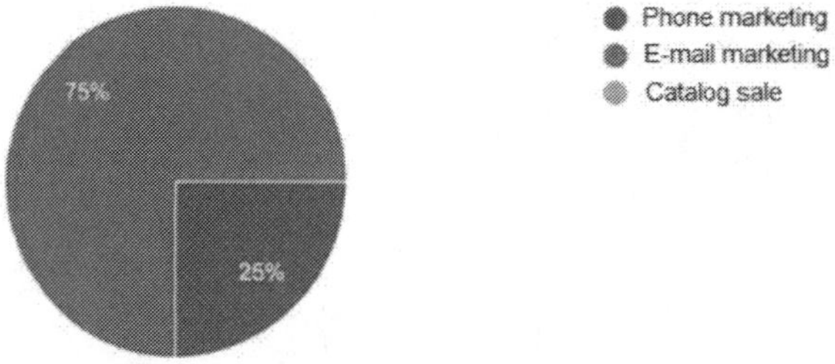

Fig. 7. The use of direct marketing as a distribution tool by logistics companies.

As for the second part of the survey (in-depth interview), the telephone interview showed that the logistics companies did not have their marketing department, dedicated for market research, attraction of cargo flows, and independent pricing policy. They delegate these functions to specialized intermediary marketing companies, which receive quite substantial amounts of money (almost half the cost of orders) for attracting freight flows and working with customers. It is assumed that instead of involving intermediary marketing companies for customer attraction, marketing departments in companies would make better output for attracting freight owners, forming and promoting the company reputation. Achieving this goal would require less financial resources than the amounts paid for the services provided by intermediary marketing companies [14-16]. Adopting these additional incentives would allow them to increase the number of costumers. This recommendation proved to be accountable to them. Therefore, the logistics companies that participated in the survey intend to take this into account.

4 Conclusions

Thus, the management of logistics companies operating in Georgia can become sophisticated by improving marketing communications, minimizing the material, financial and time costs of shipping and placing goods in the transportation and warehousing network. To achieve this goals, it is essential to:

1. Increase the special knowledge in the field of marketing among employees of logistics companies, train their managers in the principles of marketing management;
2. Create marketing departments and provide them with appropriate personnel;
3. Introduce effective marketing communication tools that will help raise the company's reputation and gain a competitive advantage;
4. Carry out systematic marketing research to ensure the reliability of the information available in the market and develop the right strategy
5. Actions related to the distribution of goods should be coordinated between the parties (a separate part of the logistics system);
6. It is necessary to choose the technological means of goods transportation correctly and rationally. The process of the cargo flow should always be coordinated between the participants in order to organize the technical side of the logistics system;
7. Optimize the workload of the logistics network, thus reduce costs and improve economic efficiency.

As a result of the above measures, it is possible to integrate the material flow system and proper management, which will ultimately ensure the successful development of logistics systems in Georgia and increase economic efficiency.

References

1. Xiangliana, C., & Huab, L. (2013). Research on e-commerce logistics system informationization in Chain. 13th COTA International Conference of Transportation Professionals (CICTP 2013). Procedia - Social and Behavioral Sciences, 96, 838 – 843.
2. Gochava, D. (2015). Logistics system efficiency and Georgia. Journal "Transport and Logistics", Tbilisi, December.
3. Sheverdashvili, E., & Totikashvili, S. (2017). The role of logistics in the development of the Georgian economy. Journal "Economy - XXI Century", Tbilisi.
4. Doborjginidze, G. (2014). The role of logistics in the development of the Georgian economy. Journal of Economics,1–2.

5. Makharadze, L. (2019). Logistics in the Georgian economy. Journal "Transport and Logistics", Tbilisi.
6. Osadze, L. (2009). The role of the logistics system in the economic activity of firms. Magazine. "Business and Management", 1, Batumi.
7. Kvlividze, N. (2020). Problems of creating a logistics cluster in Georgia. Journal "Transport and Logistics", 2, Tbilisi.
8. Brah, S.A, & Lim, H.Y. (2006). The effects of technology and TQM on the performance of logistics companies. International Journal of Physical Distribution & Logistics Management, 36(3), 192–209.
9. Laric, M.V., Lynagh, P.M., & Pikman, D. (2009). The role of promotion in supply-chain management {scm}. The Role of Promotion in Supply-Chain Management.
10. Lotfi, Z., Muriati, M., Sahran, S., & Taei, A. (2013). Information Sharing in Supply Chain Management. The 4th International Conference on Electrical Engineering and Informatics (ICEEI 2013) Procedia Technology, 11, 298 – 304.
11. Gebrüder Weiss (n.d.). https://www.gw-world.com
12. LogisticsWorld (n.d.). http://www.logisticsworld.com
13. The Ministry of Economy and Sustainable Development of Georgian. (2021) Total Cargo Transportation. http://www.economy.ge/?page=ecoreview&s=26
14. Mikhno, I., Koval, V., Korenyuk, P., Smutchak, Z., & Bozhanova, V. (2021). Modeling corporate games to increase the ecological value of entrepreneurship and innovative business. E3S Web of Conferences, 255, 01027.
15. Arsawan, I.W.E., Koval, V., Rajiani, I., Rustiarini, N.W., Supartha, W.G. and Suryantini, N.P.S. (2020). Leveraging knowledge sharing and innovation culture into SMEs sustainable competitive advantage. International Journal of Productivity and Performance Management, Vol. ahead-of-print No. ahead-of-print. https://doi.org/10.1108/IJPPM-04-2020-0192
16. Petrova, M., Koval, V., Tepavicharova, M., Zerkal, A., Radchenko, A., & Bondarchuk, N. (2020). The interaction between the human resources motivation and the commitment to the organization. Journal of Security and Sustainability Issues, 9(3), 897-907. https://doi.org/10.9770/jssi.2020.9.3(15)

The main challenges of cruise tourism and statistics of expectations in Georgia

Nino Abesadze[1*], *Rusudan* Kinkladze[2], *Maia* Giorgobiani[1], *Natalia* Robitashvili[3], and *Ketevan* Chitaladze[1]

[1]Ivane Javakishvili Tbilisi State University, Department of Economic and Social Statistics 0177 Tbilisi, Georgia
[2]Georgian Technical University, Department of Business Administration, 0171 Tbilisi, Georgia
[3]Batumi Shota Rustaveli State University, Department of Economics, 0160, Batumi, Georgia

Abstract. In parallel with global technologies and transformations, Georgia was intensively reviving and developing all types of tourism, the COVID-19 pandemic slowed the pace of rapid growth in the quantitative parameters of tourism, and then shifted tourism to a phase of economic uncertainty, leaving all types of tourism paralyzed. The main goal of the research was to identify the challenges facing cruise tourism and to develop recommendations for its development.The article discusses the importance of developing cruise tourism for the revival of tourism in Georgia. The potential of cruise tourism is characterized. According to the trends of the world, based on the annual report of the International Association of Cruise Lines for 2020, the statistics of cruises around the world are analyzed.We believe that educational projects should be started for tourism business owners, staff should be trained to give accurate recommendations regarding maritime cruises and COVID regulations, travel companies should be given the opportunity to take low-interest loans if necessary; State support for the promotion and development of tourism, including cruise tourism, is important.

1 Introduction

Along with digital technologies and transformations, all types of tourism were intensively revived and developed in Georgia, the COVID-19 pandemic slowed the rapid growth of quantitative tourism parameters, and then shifted tourism to a phase of economic uncertainty. That is why 2020, as in Georgia, as well as around the world, turned out to be a year of radical change and transformation of public relations, live contacts have been replaced by online platforms, people's lifestyles and rhythms have changed. Domestic tourism in anticipation of the revival regime is again in the dormant stage, which is why tourism in general still retains the status of passive viability. In view of all the above, the pandemic has significantly changed the parameters of Georgia's economic development, which has put business entities in a difficult situation. There is no doubt that the pandemic has had the greatest impact on the tourism sector both in terms of demand and supply [4].

* Corresponding author: nino.abesadze@tsu.ge

However, in Georgia, as in other tourism countries, the basic statistical characteristics of tourism have changed significantly, expectations for tourism revival, especially domestic tourism, exist and they Is positive. This is confirmed by various studies, including those conducted by us, according to the results of which potential visitors are waiting for the stabilization of the epidemiological situation, are in the waiting mode, but have selected attractive tourist locations and types of tourist packages. At the same time, it should be noted that it is important for both international and domestic tourism markets to create high-quality tourism products and ensure a safe corridor and affordable air and land traffic [1].

For Georgia, tourism can be said to be a key to socio-economic progress due in part to its influence and an important resource for attracting investment to the country [8]. Tourism is one of the important sectors of the Georgian economy that has great potential. It is one of the most important sources of income and employment growth [3], but it is necessary to intensify measures to promote tourism and raise awareness of the country for the post-pandemic period. Tourism revenue also increased significantly. As tourism develops, competition in the tourism industry grows. Countries are trying different types of traditional or innovative marketing PR campaigns to attract tourists. Practice shows that with good PR the results can be achieved very quickly [3].

It is known that visitors to the country come mainly by land and air. Cruise traffic capabilities are underestimated, so the scale of its use is uncertain. Clearly, increasing the intensity of cruise traffic will also contribute to the development of cruise, tours. To do this, first of all, it is necessary to pay attention to the pilot regions, where the flows of potential visitors can increase and impulse or strongly formed demands arise on cruise tours. To do this, it is necessary to ensure the diversity of cruise tourism products, even in the segment of the domestic tourism market, which should be accompanied by training to provide appropriate services to staff based on training, use of innovative methods of cruise tour business planning, etc. sh.

The main goal of the research was to identify the challenges facing cruise tourism and to develop recommendations for its development.

2 Results

Clearly, the use of digital technologies, which acquired new practical value during the Covid-19 pandemic, will play a major role in the development of tourism, including cruise tourism. Monotonous tourism packages were one of the main factors mentioned by tourists as a hindering factor in the development of tourism in Georgia. Clearly, digital transformation, as a synthesis of advanced digital technologies combining physical and digital systems, will be crucial to shaping and realizing demand for a country's diverse tourism product [5]. It can be boldly said that today, digital technologies are considered as a catalyst for the development of cruise tourism and consequently the growth of tourism in the country. Which is essential for the popularization, awareness raising and attractiveness of maritime, cruise tourism countries. Moreover, digital technologies are transforming the tourism industry and introducing new tourism products and services. Marine is a product that can become a characteristic niche of the Georgian tourism industry in terms of promotion and digital transformations, which is an inevitable process, as intensive advancement of travel companies' relationships with potential customers through online marketing, social networks and various applications. Therefore, it will be of great importance how Georgia's popularity will increase and its attractiveness coefficient in the digital world will increase.

Cruise tourism is less developed in Georgia, it can be said that it does not exist. However, it is one of the most demanding and fast growing sectors in the world. The increase in its attractiveness reinforces the trend of a systematic increase in its share in the structure of revenues from tourism.

It should be noted that cruise tourism is a very big business. So big, in fact, that according to a report by the International Cruise Lines Association, there has been a rapid growth in demand for cruise tours since 2010, for example, from 2011 to 2016, demand for cruises increased by 20.5%. Global cruise passengers reached 25.8 million in 2017, 27.2 million in 2018, and more than 32 million in 2020. The secret of the attractiveness of cruise tours is due to the fact that tourists are satisfied with the structure of the services received during the tours, with a wide range of opportunities for self-realization for leisure, offering the best and most comprehensive environment for travel lovers, etc. Such an attitude towards cruise tours was formed when analyzing the databases obtained from customer surveys. "Respondents note that more people are more open on the cruise than ever before," and that the cruise is one of the most exciting travels with families. It is the best and most comprehensive trip for those who want to travel more and more [7].

It is true that Georgia has excellent natural resources for the development of cruise tourism, but natural values alone are not enough to form new cruise-related tourism products and, moreover, to market them. It is necessary to support the state to travel companies, their wide promotion, based on various advertisements, banners, information meetings, etc.

In addition, it should be noted that cruise tourism, such as river cruises, may develop in Georgia. Large rivers of the country according to the Caspian and Black Sea basins: Mtkvari 1364 km. Including 351 km in Georgia, flows in Samtskhe-Javakheti, Shida Kartli and Kvemo Kartli, Alazani, respectively 407 km, in Georgia 390 km, flows Kakheti, Enguri, with a total length of 213 km, including 213 km in Georgia, [10] flows in Samegrelo-Zemo Svaneti, Abkhazia , Rioni with a total length of 333 km, respectively 333 km in Georgia, flows in Racha-Lechkhumi-Kvemo Svaneti and Imereti, Iori, which is 320 km long, including 183 km in Georgia, flows in Mtskheta-Mtianeti and Kakheti, Khrami is 210 km long, its length in Georgia 187 km, flows in Kvemo Kartli, etc. [9]. Can be actively used in the direction of tourism. It is true that it is the practice of mini-tours to walk on them, but the existence of perfect packages will increase the possibilities of using the rivers for tourism purposes. These types of tourism products are quite popular in Europe. Therefore, it is essential to share international experience and train tour operators on how to host a river-cruise night. The popularity of such tours is facilitated by the use of many new modernized vessels in this direction, which, given the perfection of the relevant infrastructure, has become a source of customer attraction and growth trends in cruise tourism.

According to the 2020 report of the International Cruise Lines Association, there are a total of 55 cruise lines in the world, including 98% of ocean, river and special cruises, 400 executive partners, these are major suppliers and cruise line partners, including ports and destinations, ship development, delivery and business In the direction of services. Cruise tours are served by 15,000 travel agencies and 53,000 member agencies worldwide.

It is true that Georgia has excellent natural resources for the development of cruise tourism, but natural values alone are not enough to form new cruise-related tourism products and, moreover, to market them. It is necessary to support the state to travel companies, their wide promotion, based on various advertisements, banners, information meetings, etc. According to the same report, it was engaged in cruise tourism alone and 32 million people worldwide used its services and as a diagram 1- From 2009 to 2020, the number of cruise tourists increased by an average of 5% each year.

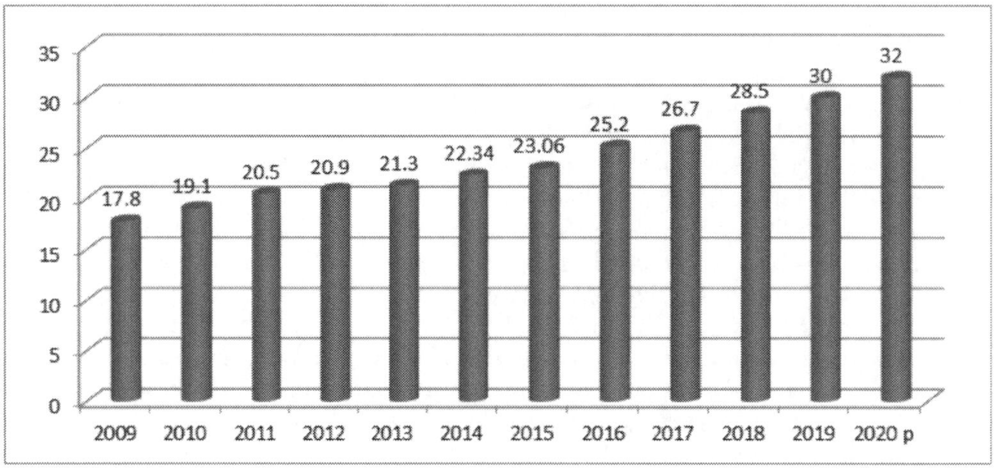

Fig. 1. Global Ocean Cruise Passengers (In Millions) [6].

You are probably wondering which locations are attractive for customers if they are interested in cruise tourism.

According to a 2020 report by the International Cruise Lines Association, 32% of all cruise lines came from the Caribbean, 17% from Mediterranean Europe, 5% from Alaska, 5% from China, and so on. It should be noted that according to the 2018 report, the list of popular locations was unconditionally led by Alaska, followed by the Caribbean, the Mediterranean, Canada, Australia and others. Sh. Regions.

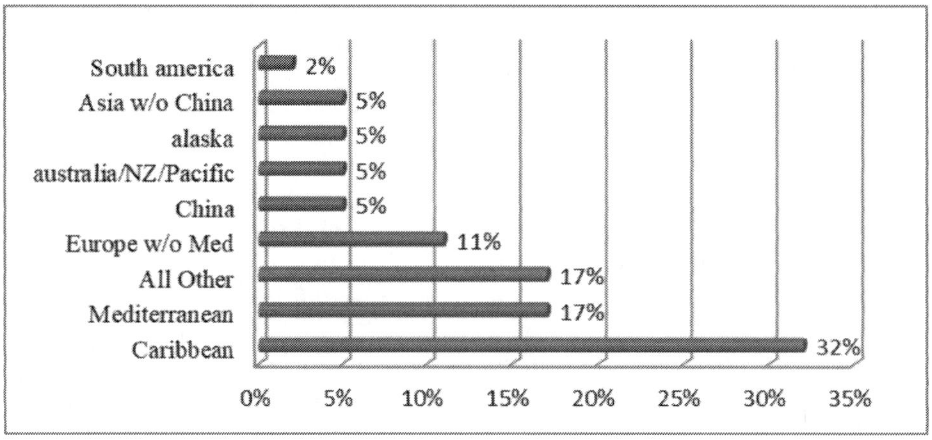

Fig. 2. Cruise Line Deployment By Region, % [6].

Where Are Passengers Coming From?

A 2020 report by the International Cruise Lines Association shows that interest rates on cruise tours were high in North America, Western Europe, Asia, Australia and New Zealand.

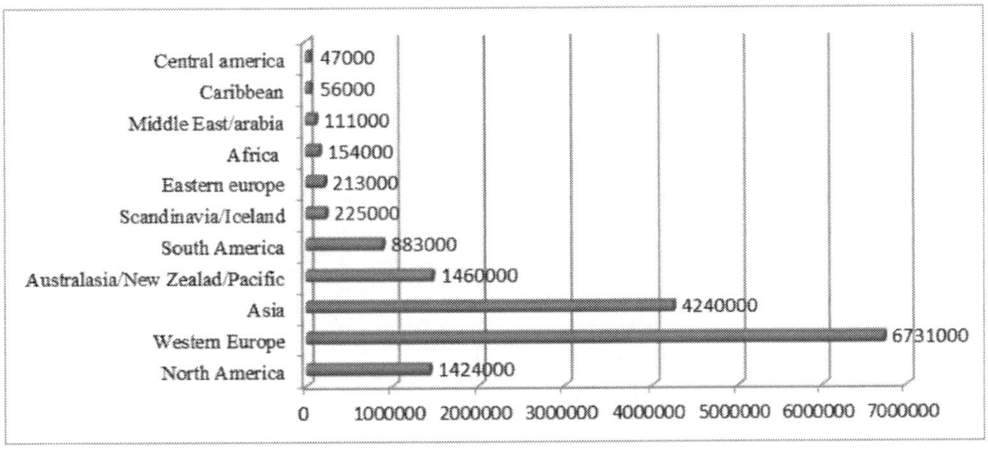

Fig. 3. Distribution of tourists by regions [6].

As it turned out, attractive cruise locations are headed by the Caribbean / Bermuda / Mexico, the list of popular destinations is Mediterranean Europe, Alaska. Canada Australia, New Zealand, North America, China, and so on. Sh.

It should be noted that according to the Cruise Lines International Association 2020 report, 82% of cruise tourism customers have expressed a desire to use the cruise service again during the next vacation.

Perhaps you are wondering what is the cost of cruise tourism per capita per capita? According to the same report, the expenses incurred by one tourist before boarding the ship amounted to 376 USD, while during the cruise itself 101 USD.

Of the countries of the former Soviet Union, most cruise tourism was developed by the Baltic states. Take, for example, the Riga-Stockholm, Stockholm-Riga cruise line, which is growing in popularity not only for its attractive package structure but also for its price. It is possible to buy a passenger ticket for one por for just 50 euros. The tourist is given the opportunity to spend the whole full day in Stockholm, with the prospect of sightseeing. This is a 2-night cruise Riga-Stockholm-Riga (or vice versa、Stockholm-Riga-Stockholm). Allowing tourists to travel with a tour guide and guide, transfer from both city center to the port, all conditions on board are created for tourists to feel comfortable and enjoy the entertainment center, restaurants and on-board bar. Additional tax cases are also included in the offer for a car tour of Stockholm with the help of a guide.

Why not organize such short-term tours from Georgia, even to Turkey or Azerbaijan?

The tourism potential of our country, the fascinating nature, the potential of water resources, the infrastructure of tourist locations create the preconditions for cruise tourism to take a worthy place in the strategy of future tourism development.

Various studies conducted in Georgia, including our own, show that the population has expectations for the development of tourism. Today, the main indicator of tourism revival is undoubtedly domestic tourism. With the Google Drive document, we conducted a survey of the population of Georgia to identify popular tourism products and attractive locations. As the results of the research showed, the population is attracted to mountain, eco, adventure tours. 28% of respondents expressed dissatisfaction with the lack of opportunities to offer cruise tours in the country, both in terms of domestic tourism, especially international outbound tourism. This indicates that as a result of proper support and transformations, it is possible to modernize Georgian tour products in accordance with new, modern requirements.

An interesting picture was revealed during the analysis of the persons accompanying the trip. As the data show, the number of respondents who want to travel with relatives and loved ones increases with age, while the share of those willing to travel with friends is higher in the case of younger age groups. It should be noted that the younger generation is most attracted to adventure tourism, and the number of people wishing for a cruise tour increased with age.

In the case of domestic tourism, as the survey showed, in 2020, 60.5% of respondents had a history of domestic tourism. A large proportion of them (58%) traveled with friends and relatives, -14% alone, and 11% of respondents with family members.

For almost 60% of domestic tourism travelers, the purpose was to relax, have fun and recreate. There were, for 14%, visiting relatives and loved ones. For 16% treatment and health improvement, for 10% professional activities. As we can see, the main goal of domestic tourism in Georgia is related to recreation. Therefore, their potential can be easily used in organizing river and sea cruises. However,

The majority of internal visitor respondents (63%) prefer the hotel as the most popular accommodation, 8% - hostel, 10% - guest house, 11% - relative's house. One of the most important factors in the effectiveness of cruise tourism is the good infrastructure tailored to the onboard customers. If the quality and price of the service are consistent, the promotion of comfortable room offerings will become a catalyst for the development of cruise tourism.

When analyzing the costs for them, 28% of respondents said that they would spend up to 500 GEL while traveling, 22% - up to 1000 GEL and 50%-up to 1500 GEL. When asked which territorial location you would choose to travel to tomorrow, 31% of respondents preferred to spend their vacation at the sea, 12% would go to visit the colorful lakes of Abudelauri, rafting in Pasanauri turned out to be attractive for 18%, Dashburi for 9%.

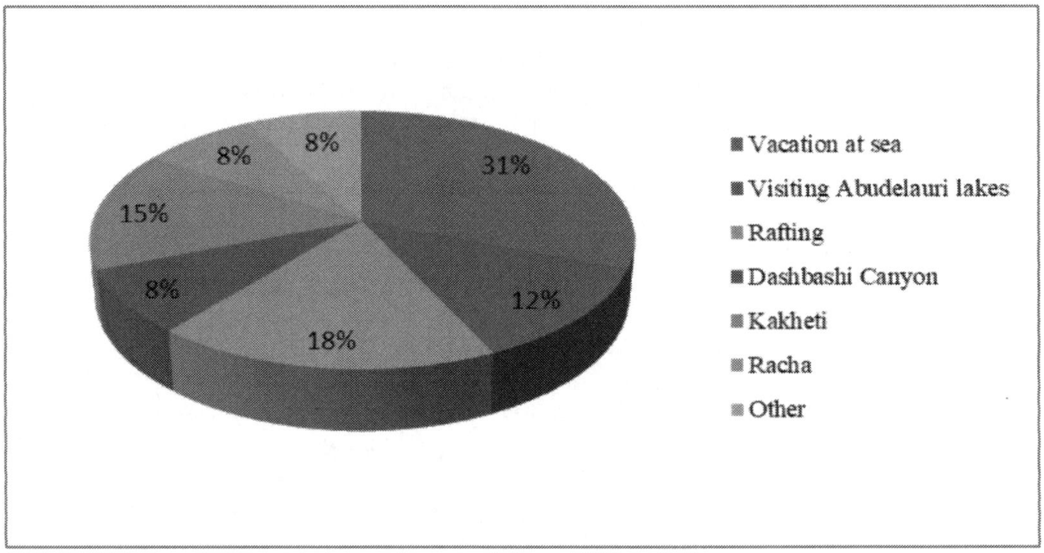

Fig. 4. Distribution of respondents according to attractive locations– %.

Water-associated locations have been named as preferred locations. This indicates that if well-packed river and sea cruise tours enter the Georgian tourism market, there will most likely be a high level of interest from Georgian residents in both international outbound and domestic tourism.

It is also a fact that the survival of tourism at this stage should be at the expense of activating domestic tourism. It is true that international visitors started arriving in Georgia on February 1, 2021, but according to official statistics, reaching the level reached by 2020 will not be possible in the near future.

That is why the importance of activating domestic tourism is growing. Along with the traditional, existing tourist packages in the tourist market, new products, which were foreign to Georgian residents, including cruise tours, must appear.

Based on all the above, it can be said that tourism in Georgia is waiting for revival. The strict regulations caused by the pandemic are slowly opening up, creating a stimulating base for domestic tourism activation with the necessary recommendations in mind. Here it will be crucial to inform the population, increase, provide a safe environment, use virtual tour applications, modern advances in digital technology, etc.

3 Conclusion

In conclusion, we can formulate the main recommendations for the development of cruise tourism in Georgia:

- Georgia should become a safe destination for visitors;
- It is necessary to provide safe and secure cruise traffic along with affordable and safe air and land traffic, which is one of the catalysts for increasing the number of tourists;
- In order to activate cruise tourism, it is necessary to increase the interest in domestic tourism;
- Educational projects should be started for tourism business owners, staff should be trained to give accurate recommendations regarding cruises and COVID regulations.
- Travel companies, if necessary, should be given the opportunity to take low-interest loans;
- It is important to support the state for the promotion and development of tourism, including cruise tourism;
- It is necessary to have thematic meetings (including online mode), news exhibitions, videos, information banners, virtual tours, applications with built-in voice guide, etc. sh.

References

1. Abesadze, N., Abesadze, O., & Amanova, L. (2020). Trends Of The Attractiveness Of Azerbaijan Tourists In Georgia. Economic and Social Development. Book of Proceedings, 1(4), 55th International Scientific Conference on Economic and Social Development. Baku.
2. Abesadze, N., Abesadze, O., & Amanova, L. (2020). Trends In The Growth Of Tourist Flows From Azerbaijan To Georgia. The Journal Of Economic Sciences: Theory And Practice, **77**(1).
3. Abesadze, N., Kinkladze, R., Paresashvili, N., Cjitaladze, K., & Meoshvili, S. (2020). Trends of increases of Georgia's attractiveness for EU Countries. Brussels, Belgium.
4. Abesadze, O. (2020). Post-pandemic tourism expectations and development trends in Georgia. Economics, 10-12.
5. Amanova, L., Abesadze, N., & Abesadze, O. (2021). The Importance of digital technologies in the revival of tourism in Georgia and Azerbaijan. Baku. Azerbaijan.
6. Cruise Lines International Association. (2020). State Of The Cruise 2020 Industry Outlook. https://cruising.org
7. Cruise Lines International Association (n.d.). Travel agent cruise industry outlook. https://cruising.org
8. Tikishvili, M., Giorgobiani, M., & Shonia, Z. (2014). Ecotourism and its infrastructure development prospects based on statistical calculations in EXCEl. Batumi.
9. Reading Ladder (n.d.). http://reading.ge/ka/books-by-level/192-saqartvelos-mdinareebi
10. Saqartveloi (2015). Rivers of Georgia. https://saqartveloi.blogspot.com/

The role of sustainable transport system in the competitiveness of agri-food products

Mariam Saghareishvili[1*]

[1]Ivane Javakhishvili Tbilisi State University, 1 Ilia Chavchavadze Avenue, Tbilisi, Georgia

Abstract. The article discusses the role of sustainable transport systems in competitiveness of Georgian agri-food products. Georgian agri-food products are less competitive on international markets. The country has FTAs with its partners that should make Georgian goods more competitive, at least in terms of price. Transport and logistics is one of the main parts of export, therefore, they have an impact on competitiveness of exported goods. Georgia trades with its neighbour countries, with the EU, China, etc. The EU and China represent huge markets and Georgian government targets to increase export in both directions, but there is a long distance between Georgia and its target markets, therefore it needs a developed transport and logistics system. On the other hand, portions of exported Georgian agri-food products are primary production, which have a shorter lifetime period, that's why they should be transported in a short period of time, with specific machines. Without development of the transport system it is impossible to achieve sustainable export of agri-food products into Georgia's target markets. Investments in the system, with tangible strategy and holistic approach can create a sustainable transport system that will increase competitiveness of Georgian agri-food products in international markets.

1 Introduction

Agricultural sector in Georgia faces many challenges, one of which is agri-food competitiveness on international markets. Georgian government implements programs and projects to support the sector. Trade liberalization is one of the ways of increasing competitiveness of a country's goods and services.

In order to liberalize trade, countries apply for free trade agreements. Georgia also pursues a policy of trade liberalization, the country shares the principles of the World Trade Organization and at the same time, has already signed free trade agreements with many countries. Georgian companies trade with EU countries, the European Free Trade Association (EFTA) countries, China, Hong Kong, Turkey, the Russian Federation, Azerbaijan, Armenia, Ukraine, Moldova, Kazakhstan, Uzbekistan and Turkey within the framework of the free trade regime.

The free trade agreement with these countries includes the simplified circulation of both products and, in some cases, services between the countries.

* Corresponding author: m.saghareishvili@iset.ge

The terms of the agreement, of course, apply to agricultural products. Free trade ensures the free movement of goods and services between countries. In addition, free trade ensures that countries share knowledge, experience and modern approaches and technologies. By sharing modern approaches and technological knowledge it is possible to increase productivity, production and competitiveness on different markets.

Competitiveness of agri-food products depends on several factors. For example, costs of production, quality of the products which can be achieved by following international standards of food production, developed logistics and supply chain of the products.

Development of competitiveness of agricultural products is impossible without a sustainable supply chain and transport system has an important role in it. Sustainable transport systems can positively contribute to competitiveness of agri-food products on international markets. Sustainable transport systems are able to provide efficient mobility of goods and services across and outside the country.

Improvement of transport systems can be achieved by increased investments in the sector. Investments lead to increased transport capacity, higher efficiency and quality service. Benefits of developed transport systems are low transport costs, short transit time, ability to expand a business, increased productivity, all of which results in increased competitiveness and economic growth [1].

This article assesses the impact of sustainable transport systems on competitiveness of agri-food products on international markets. The article reviews the role of the transport system in competitiveness in different countries. Also, it discusses how Georgian transport system is developed and how it contributes to competitiveness of agri-food products.

2 Literature review

Competitiveness itself is defined by different economists and institutions. Most common definition of competitiveness is by Michael Porter, who defines competitiveness as national productivity, providing a higher standard of living and employment [2].

Sustainable transport system is a capacity that ensures mobility of the society with the least damageable manner to the environment and does not decrease mobily of the future generation [3].

Many countries recognize the role of a sustainable transport system in an economy and competitiveness. For example, researchers in Lithuania analyzed the effect of a sustainable transport system on the competitiveness of the national economy. Lithuania's economy benefited from transportation through improved mobility and improved quality of service. However, development of transport systems has its drawbacks which mainly is associated with increased noise and air pollution, and increased number of traffic accidents. Considering the drawbacks, overall, allocation of investment in the transport system gave Lithuania a chance to increase economic competitiveness and achieve rapid economic growth [1].

Investments in transport infrastructure improve accessibility of the regions however, its impact on competitiveness is unclear [4]. Transport system consists of three markets: trip, transport and traffic markets. Trip market consists of distribution of location and time. In the transport market, vehicles represent the supply side and trip patterns represent the demand side. In the traffic market, the supply side consists of infrastructure and traffic control systems. All three markets must work properly to have a sustained transport system and improve accessibility. Other factors that affect accessibility are mobility, integrated transport system, affordability of transport, transport network connectivity, management of roadways.

The article "Transport and agricultural productivity: a cross-national analysis" discusses the role of transport infrastructure in agricultural development.

To measure the link between the two, authors use data for transport infrastructure meaning total road network (highways, main and secondary roads, etc.); agricultural production; agricultural inputs, such as land and machinery; and education. Empirical analyses showed that comparing two countries with the same education level and agricultural inputs, a country with a better transport system can produce more agricultural goods. This is a direct effect of a sustainable transport system on agricultural productivity, while there also is an indirect effect which implies that accumulation of a sustainable transport system means accumulation of infrastructure in the country, that can increase countries productivity and cause higher economic growth [5].

Sustainable transports and logistics not only increase competitiveness of agricultural products but also increase food security, which became even more important during the Pandemic. Avoiding food waste and efficient allocation of products is vital in periods of lockdowns and limited transportation [6].

3 Methodology

To identify the role of sustainable transport system in competitiveness of Georgian agri-food products, the article discusses the following pillars:

- Types of transport in Georgia, especially for export purposes;
- Currents state of transport system in Georgia;
- Competitiveness of agri-food products in Georgia;
- Factors that hinder development of transport system in Georgia;
- The role of sustainable transport systems in competitiveness of agri-food products.

To outline agri-food export tendency and most common means of transport in Georgia's export, the data of the national statistics office of Georgia was reviewed. To evaluate Georgia's current situations in terms of transport system, the paper also discusses Georgia's positions in international rankings related to transports and logistics.

4 Analysis

Agricultural sector has been prioritized by Georgian government since 2012 and a lot of programs and projects have been implemented to increase productivity and competitiveness of the sector [7]. One of the aims of the Ministry of Environmental Protection and Agriculture of Georgia is to increase competitiveness of Georgian agri-food products. Moreover, the government aims to develop agricultural value chains, part of which is transport and logistics [7]. Georgian agriculture is characterized by production of primary goods rather than secondary goods. This means that low value added products are produced in a country, rather than high value added products. Primary products have low prices, but the prices still are not competitive on international markets.

According to the National Statistics Office of Georgia, total export of food products amounted to 317 million USD in 2020. On average the share of agri-food export in total export was 9.3% in 2020, which is higher than the same variables in 2018 and 2019. Table 1 shows the top ten agri-food products exported from Georgia, where fruits, vegetables, wheat represent primary products.

Development of supply chains and value chains have vital importance for increasing competitiveness of agro-food products [8]. One of the main parts of the supply chain are transport and logistics. This part of the chain is important not only for international markets but for mobility of the goods and services within a country.

https://doi.org/10.1051/matecconf/202133901007

Table 1. Top ten food products exported from Georgia [16].

Products	Value (million USD)
Wine	210.3
Fruit	144.5
Spirits	132.2
Mineral waters	116.6
Soft drinks	31.6
Vegetables	10.3
Beer	5.1
Sugar	2.2
Vegetable oil	0.5
Wheat	0.01

Georgia's transport system consists of road, railways, sea and air transport and pipelines [9]. Transport infrastructure is improving every year, the government implements investment projects to modernize the infrastructure. Main sources of transport for food products are sea and road transport. According to the National Statistics office of Georgia, in 2020 \$3.3 billion was exported from Georgia, of which 47% was exported by sea and 42% by road transport, only 11% of total export was through air transport and railway (Fig. 1).

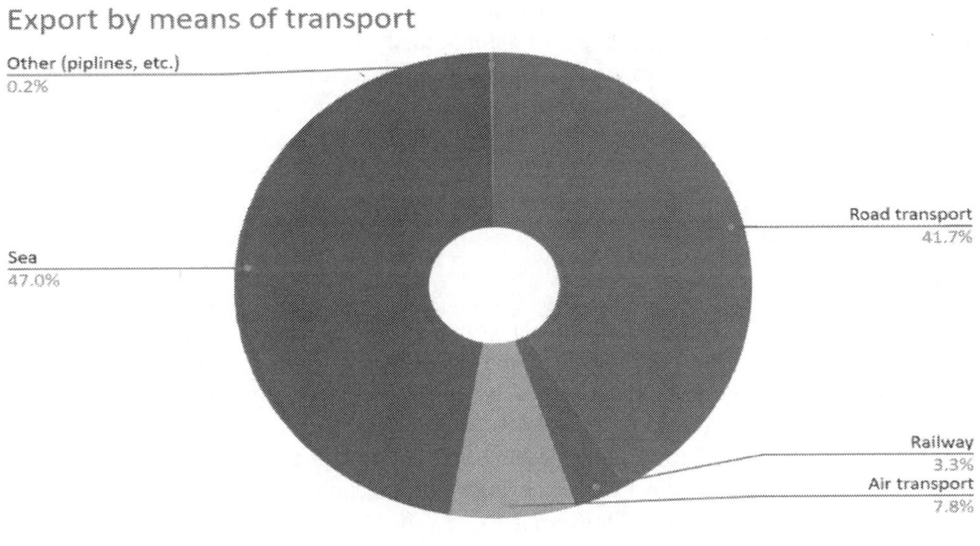

Fig 1. Export by means of transport [12].

Usage of sea transport has been increasing every year since 2016. It increased by 5% in 2020. Usage of road transport was also increasing from 2016 until 2020, however it sharply declined by 26% in 2020 compared to the previous year.

Usage of air transport and railways were also increasing from 2016 until 2020, however they declined in 2020 by 0.3% and 25%, respectively. Decrease of road, air and railway transport can be explained by COVID-19 outbreak in the world, which caused lockdowns in most of the countries in the world and created delays in supply chains. Representatives of the agricultural sector struggled with pandemic restrictions, according to the survey "COVID-19 impact on Georgian farmers and agriculture" 55% of surveyed farmers face challenges in transportation due to COVID-19 restrictions [10].

In the World Bank Logistics Performance Index Georgia ranked 119 in 2018. The index evaluates supply chain service delivery, which includes six indicators - customs, infrastructure, international shipments, service quality, tracking and tracing, timeliness. Among the indicators, Georgia has the highest rank in customs, the rest indicators have the following ranking [11]: Customs (95); Infrastructure (102); Timeliness (105); International shipments (124); logistics quality and competence (132); tracking and tracing (139).

Overall, Georgia's logistics performance is quite low with underdeveloped infrastructure, delays in delivery and quality of logistics service. Competitiveness of Georgian goods and services will improve if its logistics will become more predictable and reliable, especially, shipment delivery.

Georgia ranked 74 in the Global Competitiveness Index by the World Economic Forum in 2019 [12]. The index consists of 11 pillars, one of which is infrastructure. The pillar consists of transport and utility infrastructure. Transports infrastructure ranks 83 and it evaluates the country's performance in road connectivity, quality of read infrastructure, railroad density, efficiency of train services, airport connectivity, efficiency of air transport services, liner shipping connectivity, efficiency of seaport services. Within the pillar efficiency of train services has the highest rank (43), while liner shipping connectivity - the worst (100). Georgia's performance in the international rankings show that transport infrastructure has to be developed further in the country.

Georgia targets to increase its export to European Union. Having a Free Trade Agreement with the EU is a good basis for it. However, Georgian agri-food products are not competitive on the EU market. The value chain analyses of several agricultural products with high export potential in European Union admit that transportation has one of the main roles in competitiveness of agricultural products. As a lot of goods in Georgian agri-food exports represent primary goods, their transportation requires special containers and freezing machines.

Hiring this type of machine is costly. Insurance of the cargo is an important part of the export, however, it increases the costs as well. According to the study, transportation of goods in Europe costs about 4000-5000 euro. This cost is quite high for primary food producers, that is why they prefer to export their products in traditional markets, such as Russia, Ukraine, Turkey, Azerbaijan and Armenia. If producers will pay these costs, their costs will increase and so will the price of the product. This will make Georgian products less competitive, because price is one of the factors determining competitiveness of the product [13].

Georgia also has a free trade agreement with China that successfully contributed to increased export, especially export of wine. Further expansion of export and more intensive trade can be achieved by the "Belt and Road" initiative from Chinese government. Infrastructure projects are being implemented in the framework of the initiative, which will lead to a deepening of the relationship between Georgia and the countries involved in the route. This project can also lead to development of a unified trading system in the South Caucasus which will improve competitiveness of the region and its trade [14].

5 Conclusions

Georgian agri-food products have a potential to become more competitive on international markets with development of a sustainable transport system. Signing free trade agreements with the European Union and countries like China, show that Georgia's target export markets have a big population and variety of consumer preferences, this means that Georgian government targets expansion of export, including export of agri-food products. As Georgian agri-food export consists of primary products, they must be transported in a short period of time, which is impossible without sustainable transport infrastructure, considering the fact that main means of transport for Georgian food export are sea and road transport.

Development of a sustainable transport system links environmental protection, economic efficiency and social aspects such as standard of living and quality of life.

Infrastructure is being developed in the country, investments are made in this direction, however sustainable transport infrastructure can be created with a holistic view and a tangible strategy, considering the country's export potential and target markets. To develop transport infrastructure Investment should be made in the sector. Investments in sustainable transport system are necessary because they lead to [15]:

- Geographical specialization that leads to increased productivity and comparative advantage of the country's products
- Transportation of goods over longer distances, which facilitates production in economies of scale
- Increased competition and increased variety of products for consumers
- Increased value of land

Development of a sustainable transport system requires lots of investments, however in the long-run it enables exported goods to become more competitive on international markets. Competitiveness of agri-food products is not defined only by transport and logistics, but without development of this system it is difficult to achieve sustainability of the export, especially of agricultural products.

References

1. Mačiulis, A., Vasiliauskas, A.Vasilis, & Jakubauskas, G. (2010). The impact of transport on the competitiveness of national economy. Taylor & Francis online, Transport, 24(2), 93-99. https://www.tandfonline.com/doi/pdf/10.3846/1648-4142.2009.24.93-99
2. Siudek, T., & Zawojska., A. (2014). Competitiveness in the economic concepts, theories and empirical research. Oeconomia, 13(1), 91–108.
3. Rodrigue,J.P. (2006). The geography of transport systems. Transportation, Sustainability and Decarbonization. https://transportgeography.org/contents/chapter4/transportation-sustainability-decarbonization/
4. Kiel, J., Smith, R., & Ubbels, B. (2013). The impact of transport investments on competitiveness. 41st European Transport Conference. https://core.ac.uk/download/pdf/81134137.pdf
5. Lui, S. (2017). Transport and agricultural productivity: a cross-national analysis. Research on Modern Higher Education 2, 01012. https://www.researchgate.net/publication/314219295_Transport_and_agricultural_productivity_A_cross-national_analysis

6. Irigoyen, J.L. (2014). To feed the future, let's make logistics and transport sustainable. World bank blogs. https://blogs.worldbank.org/transport/feed-future-let-s-make-logistics-and-transport-sustainable
7. Strategy for agricultural development in Georgia 2015-2020. (2021). The ministry of environmental protection and agriculture of Georgia. https://mepa.gov.ge/En/PublicInformation/30/
8. PMC research center. (2018). Peach, Tangerine, raspberry value chain analysis.
9. Asian Development Bank. (2014). Georgia Transport Sector Assessment, Strategy, and Road Map. https://www.adb.org/sites/default/files/institutional-document/34108/files/georgia-transport-assessment-strategy-road-map.pdf
10. Georgian Farmers' Association. (2020). COVID-19 impact on Georgian farmers and agriculture. https://gfa.org.ge/wp-content/uploads/2020/04/GFA_COVID-19IMPACT_ENG_compressed_compressed-min_reduce_reduce-1.pdf
11. The world bank. (2018). Logistics performance index and its indicators. https://openknowledge.worldbank.org/bitstream/handle/10986/29971/LPI2018.pdf
12. The world economic forum. (2019). The Global Competitiveness Report 2019. http://www3.weforum.org/docs/WEF_TheGlobalCompetitivenessReport2019.pdf
13. PMC research center. (2019). Analysis of agricultural sector: Value chain and export potential. https://www.research.pmcg-i.com/policypapers_file/8f955d78c80e5544e.pdf
14. Zabakhidze, M., et al. (2019). Connectivity, Trade and Financial Integration of the South Caucasus Via the Belt and Road Initiative (BRI). Caucasus analytical digest, 111. https://www.research.pmcg-i.com/policypapers_file/6a7b5dd7a8a1a994b.pdf
15. Rodrigue, J.P., & Notteboom, T. (2006). The geography of transport systems. Transport and economic development. https://transportgeography.org/contents/chapter3/transportation-and-economic-development/
16. National statistics office of Georgia. (2021). https://www.geostat.ge/ka

The role of maritime logistics and supply chain in stimulating the export of agricultural products

Nino Sagareishvili [1*]

[1]Ivane Javakhishvili Tbilisi State University, 1 Ilia Chavchavadze Avenue, Tbilisi, Georgia

Abstract. The article discusses the role of sustainable transport systems in competitiveness of Georgian agri-food products. Georgian agri-food products are less competitive on international markets. The country has FTAs with its partners that should make Georgian goods more competitive, at least in terms of price. Transport and logistics is one of the main parts of export, therefore, they have an impact on competitiveness of exported goods. Georgia trades with its neighbour countries, with the EU, China, etc. The EU and China represent huge markets and Georgian government targets to increase export in both directions, but there is a long distance between Georgia and its target markets, therefore it needs a developed transport and logistics system. On the other hand, portions of exported Georgian agri-food products are primary production, which have a shorter lifetime period, that's why they should be transported in a short period of time, with specific machines. Without development of the transport system it is impossible to achieve sustainable export of agri-food products into Georgia's target markets. Investments in the system, with tangible strategy and holistic approach can create a sustainable transport system that will increase competitiveness of Georgian agri-food products in international markets.

1 Introduction

Development of international trade and globalized economies highly depend on the efficiency and sustainability of logistic services. There are many challenges during pandemic for international logistics and maritime industries. However, it should be mentioned that besides the challenges development of maritime industry becomes a great business opportunity [1].

Nowadays, during the globalization process can be seen that no nation is self-sufficient. It means that there are flows of goods and services, labor, technology, information and etc. [2].

The global economies are characterized by integrated full package services, such as: logistics, finance, production, retail and distribution. Dr. Jean-Paul Rodrigue in his article: "Globalization and International Trade" (2020) discusses the main challenges and

*Corresponding author: n.sagareishvili@iset.ge

perspective of improving transport and logistics. He states very important point that development of International trade is highly related to effective management of all elements of supply [2].

While countries are concentrating on implementation of export promotion strategies for agricultural and non-agricultural products, they have to take into account the importance of maritime logistics and effective supply chain. There is no an exact definition of maritime logistics, but Dong-Wook Song and Photis Panayides in their article: "Maritime Logistics: A New Definition" (2015) explain that maritime transport include shipping and port and is used to transport goods and passengers between different seaports by sea [3]. Logistics is defined as the overall process that a good needs to go through from the producer to get to the final consumer. These processes include: acquire, store, and transportation [4]. The management of Supply chain includes the integration of all parts of business operations across the supply chain. The main idea to discuss maritime logistics and supply chain together is to understand all key elements that are covered to get to the final consumer [3].

2 Literature review

Azmat Gani discusses very interesting studies about linkages of logistics and trade [1]:

1. Logistics for trade include wide range of processes to get to the final consumer (or country);
2. Goods that are crossing the border go through the administrative procedures;
3. There are needed wide range of international shipment operations [5];
4. There is needed effective supply chain management;
5. All processes are linked to each other and depend highly on effective management and planning system.

James Anderson and Douglas Marcouiller discussed main obstacles that international trade may face [6]:

1. Corruption
2. Imperfect contracting processes
4. Ineffective management system
5. Problems with maritime logistics
6. Trade barriers.

One of the determinant of trade can be assumed to be a shipping connectivity, discussed by Marco Fugazza and Jan Hoffmann. Maritime transport is one of the main transport services that is used during export of different types of good. The article underlines the importance of effective management of maritime logistics. When the process is implemented correctly, taking into account all administrative support, trade becomes easier between countries [7].

United Nations Conference on Trade and Development analysis Liner Shipping Bilateral Connectivity Index (LSBCI) with a data of different countries. The LSBCI includes the following 5 components [7].

1. The number of transshipments required to transport goods from country 1 to country 2;
2. The number of direct connections (for both countries);
3. The geometric mean of the number of direct connections;
4. The level of competition on services;
5. The size of the largest ships on the smallest route.

LSBCI is very important to control and monitor efficient transport services that are highly correlated with economic development. Maritime logistics is one of the main element to evaluate the competitiveness of the country in the region and in the world [7].

Georgia has favorable geographical location and has the potential to become a logistic hub of the Caucasus and Central Asia. It increases competitiveness of Georgia.

Nowadays, Georgia highly depends on import of agricultural and non-agricultural products. Mikheil Tokmazishvili in his article: "New Challenges of Transportation of Goods through the Transport Corridor of Georgia" (2018), discusses the problems of the Georgian transport corridor, dynamics of international cargo transportation, reasons for change and prospects. It is mentioned that the further development of the Georgian transit corridor is largely depended on a "one belt - one way" strategy initiated by China [8].

The article: "Georgia as the Sea Gate of the Middle Corridor" (2019), discusses that according to 2017 data, the volume of containers (TEU) shipped between Europe and Asia is about 22 million. 15 million TEUs are shipped from China to Europe and 7 million from Europe to China. 97% of the volume (in tons) of these cargoes and 70% of the value of these cargoes are transported by sea. Only 2%of cargo volume are transported by air [9].

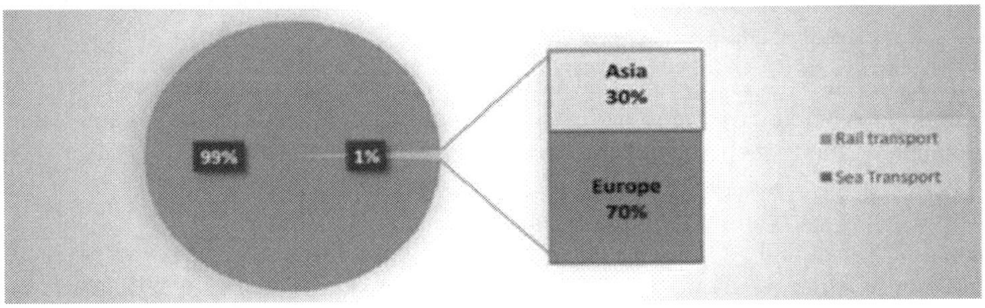

Fig. 1. Asia - Europe Container Shipping (2017) [9].

Figure 1 shows Shipping statistics during 2017. Sea transport is used in 99% while shipping goods and just 1% uses air transportation. 30% is transportation is from Asia and 70% from Europe [9].

The article underlines that the main challenge for Georgia is intermodal transportation. Intermodal transportation is defined as "using two (or more) carriers, to transport goods (freight) from shipper to consignee" [10]. Intermodal transportation needs special standardized containers of cargo on trucks, freight trains, and ships [10].

The following information are needed to find whether the intermodal transportation is good for agricultural goods or not [10]. It is most suitable when:

1. The country transports intermediate or finished goods and load of them is less than 25 tons.
2. Theree is a long distance between countries.
3. The cargos' value is intermediate. Cargo with high values are sent directly by air cargo (railway transportation is also possible).
4. There is a need to continuously transport goods to same location.

Intermodal transportation is important for the connectivity of the Caspian and Black Sea and for the efficiency of land container transportation (Hinterland transportation). The development of intermodal transport is an important factor in increasing the efficiency of ports and the functioning of the transport corridor as a unified logistics' network [9].

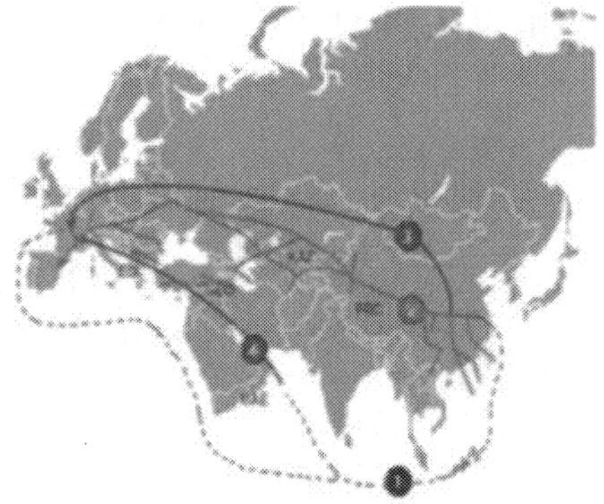

Types of transport		Time needed for transportation
Sea transport		36 Days
Air transport		14 Days
Railway		2 Days
Sea-air transport		25 Days

Fig. 2. Asia Europe Alternative Transport Routes [9].

From the Figure 2 we can find that Georgia is the shortest maritime union for Central Asia through the Middle Corridor and is therefore one of the most important central country in the corridor. Due to its geographical advantages, Georgia can become not only a regional logistics hub in the Caucasus, but also an important Central Asian transport route and an alternative transit corridor connecting China and Europe in the long term.

Georgia's transit and logistics potential plays an important role in the development of its economy and Georgia's integration into the global market. To achieve this goal Georgia requires the systematic development of transport and logistics infrastructure, the introduction of new technologies and the application of international management practices in transport and logistics [9].

Giorgi Danelia discusses opportunities of Georgia to increase transit cargo turnover. This highly depends on regional counties' economic and trade activities. Georgia is positioned as an importer country of agricultural and non-agricultural goods, but it has a potential to become a logistic hub and concentrate on increasing domestic production, promote export, especially export of agricultural goods [11].

Agricultural sector is one of priority sector for Georgian government. Country has a potential to produce high quality agricultural products that will become competitive and take place on international market [12, 13].

Main Challenge is to produce high quality agricultural products with low cost and set competitive price. Georgia's agricultural sector is developing nowadays. Government tries to increase domestic production and substitute import, so that the market will be depended on locally produced goods. Import substitution and export promotion are discussed as two strategies implemented to speed economic growth, increase employment rate and make a country competitive on international market [14]. State programs that facilitate exports stimulation can be considered [14]:

1. Subsidizing;
2. Removal of barriers for international trade;
3. Attracting financial resources.
4. Export lending and insurance.

Covid-19 became one of the main obstacle of developing agricultural sector, especially it became difficult to increase domestic production and promote export.

Georgian Farmers' Association underlines the importance of self-sufficiency coefficient. High coefficient means that a country has high potential to satisfy domestic demand, low coefficient means that a country is depended on imported goods. From data 2018-2020 the lowest self-sufficiency coefficient has wheat and meat, high coefficient has: eggs, potatoes, grape. It means that Georgia has a potential to increase domestic production of last mentioned goods and promote export. Also, we have to think if is it possible to use maritime logistics while exporting these goods? [15, 16. Nowadays it is difficult to say, because it needs more analysis and Georgia has to work more to increase effectiveness of all logistical processes.

3 Methodology

To evaluate the role of maritime logistics and supply chain management to promote export of agricultural products, the following methodologies are used: methods of analysis and synthesis, quantitative and qualitative, groupings, data analysis and graphical representation of the research.

1. There were done a questionnaire to find the effect of Covid-19 on development of agricultural sector.
2. The questionnaire gives a data that is graphically analyzed.
3. Also, the research contains the analysis of connection of agricultural sector and maritime logistics.

During the last decade demand on healthy agricultural goods increases. People are motivated to eat healthy high quality goods that are produced domestically. In the modern world with technological progress, a lot of investment are made in delivery services. Also, globalization and trade liberalization processes made countries to start developing of food distribution and logistics systems.

In the article: "The Impact of the Free Trade Policy on the Production of Local Agricultural Goods", the author underlines the advantages and disadvantages of free trade. As for advantages, we can discuss elimination of trade barriers, such as tarrifs. But, there are non-tariff barriers that can be discussed as one of the main obstacles for Georgian agricultural products to become competitive on international market [17].

Such kind of non-tariff barriers are sanitary and phytosanitary measures. Low quality of products make them less attractive not only for international consumers, but for local consumers as well. New technologies and high skilled labor force can increase agricultural productivity [17].

The article: 'Transport and agricultural productivity: A cross-national analysis", underlines that logistics and transport infrastructure are motivational factors to develop agriculture sector, because when the logistic process is effective, productivity increases. Countries which have such kind of benefit can specialize in the production of agricultural goods, get comparative advantage and promote export [18].

4 Analysis

Data is collected from the web-page of the United Nations Conference on Trade and Development. All information is discussed theoretically and graphically.

During 2019 export structure by product groups are the following [19]:

1. All food items – 23%
2. Ores and metals - 20%
3. Manufacturad goods – 54%
4. other – 3 %

Analyzing UNCTAD statistical data, we can see that there is +3.4 % growth of transport services export and +82.3% Fleet growth rate in2019 [19].

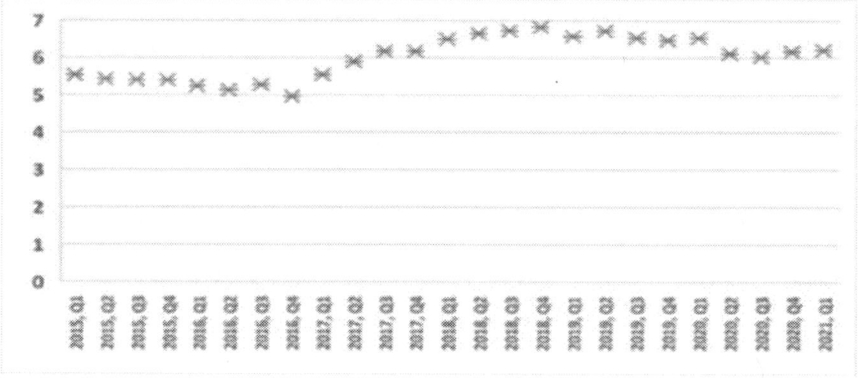

Fig. 3. Liner Shipping Bilateral Connectivity Index, Georgia [19].

Also, it is interesting to calculate LSBCI for Georgia. Analyzing data from 2015 – 2020, Georgia had decreasing tendency of LSBCI during 2015 – 2016. After that Georgia has increasing tendency till 2020. From 2020 Q1 to 2020 Q2 LSBCI decreased by 0.07%. In the Q1 of 2021, Liner Shipping Bilateral Connectivity Index is 6.21 that is higher it was in 2015. It should be mentioned that the highest level was achieved in the Q4 in 2018 [20].

Analyzing World Bank statistics, we can find that world export has a reducing tendency from 2018. Maximum amount of export in the world was reached in 2018 ($20 trillion), in 2020 total export was approximately $17 trillion. Also, figure 6 represents main food export groups in the world [21].

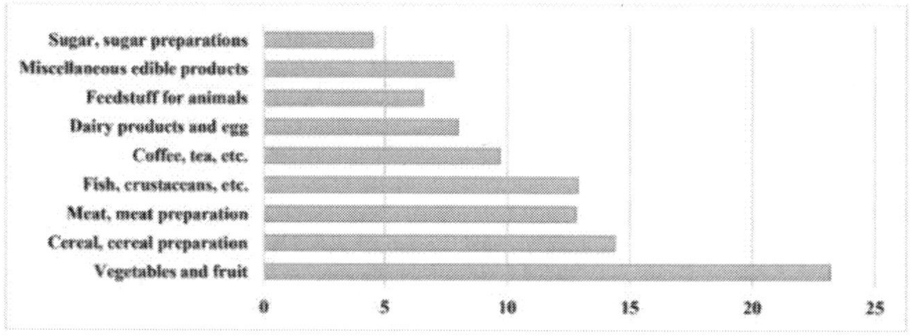

Fig. 4. Structure of food export, 2017.

According to questionnaire and the data collected by Georgian Farmers' Association, Covid-19 has a huge effect on the production of agricultural goods. These effects are discussed with different directions [15]:

1. Changes in prices (prices are increased highly during pandemic);
2. Difficulties to buy high quality and healthy products, because of Lockdown situation;
3. Increase of input prices;
4. Challenges of maritime logistics, because of regulations on economic activities;
5. People are not willing to work during pandemic;
6. First reason is Lockdown situation and regulations on transport;
7. Second reason is risk of health.

Officially 62% from the farmers and agricultural business representatives are producing goods for domestic market, only 17% export their products [15].

Also, analyzing the questionnaire, we get that 55% of agricultural business representatives have problem with supply chain. The respondents listed the main reasons of this:

1. Imported agricultural goods became more expensive because of devaluation of GEL;
2. Challenges to attract funds during pandemic;
3. Challenges of sea transportation and administrative barriers during pandemic;
4. Deficit of domestically produced intermediate and final goods (inputs);
5. Transport regulation locally and internationally.

From the data and questionnaire analysis we can conclude that political support is very important not only to developing of agricultural sector, but also develop food shipment. Government form different countries have to be ready to open borders and eliminate barriers to stimulate export and develop logistical services [22-25]. Also, government support is very important while discussing administrative barriers of shipment.

To develop shipping infrastructure, it is very important to attract funds, to implement effective regulation system, use new technology, manage supply chain, train employees, increase quality of agricultural goods and etc.

5 Conclusions

Based on analysis main conclusions are made:

1. Georgia has undeveloped shipment infrastructure that is a barrier to enter the international market
2. Agricultural sector is main priority of the government, but still there is not effective, modern standards and processes of supply chain management
3. Maritime logistics in one of the main obstacle to promote agricultural good's export
4. There is not produced high quality agricultural goods in Georgia and it makes country less competitive on the international market
5. Unskilled labour force

Is would be beneficial for the country to:

1. Improve the maritime logistics that will help not only to promote agricultural goods, but will be the real proof for Georgia to become a logistic hub of the region.
2. Increase intermodal service efficiency that will be helpful to increase export of agricultural goods.
3. Manage supply chain effectively.
4. Eliminate trade barriers for agricultural good.
5. Develop strategic partnership with regional countries and also implement effective strategy to take place on developed countries' markets.

References

1. Gani, A. (2017). The Logistics Performance Effect in International Trade. The Asian Journal of Shipping and Logistics, 33(4), 279-288. https://www.sciencedirect.com/science/article/pii/S2092521217300688
2. Rodrigue, J. (2020). Globalization and International Trade. The Geography of Transport Systems. https://transportgeography.org/contents/chapter7/globalization-international-trade/
3. Song, D., & Panayides, P. (2015). Maritime Logistics: A New Definition. https://www.koganpage.com/article/maritime-logistics-a-new-definition
4. Investopedia (2021). https://www.investopedia.com/terms/l/logistics.asp
5. Korinek, J., & Sourdin, P. "To what extent are high-quality logistics services trade facilitating". OECD Trade Policy Papers, 108, OECD Publishing.
6. Anderson, J.E., & Marcouiller, D. (2002). Insecurity and the pattern of trade: an empirical investigation. Review of Economics and Statistics, 342-352.

7. Fugazza, M., & Hoffmann, J. (2017). Liner shipping connectivity as determinant of trade. Journal of Shipping and Trade, 2(1).
8. Tokmazishvili, M. (2018). New Challenges of Transportation of Goods through the Transport Corridor of Georgia. http://eprints.tsu.ge
9. Doborjginidze, G. (2019). Georgia as the Sea Gate of the Middle Corridor. https://forbes.ge/saqarthvelo-rogorts-shua-de/
10. American Group. (2017). What is Intermodal Transportation and When is it the Best Choice? https://www.shipag.com
11. Danelia, G. (2019). Georgia's Container Market and The Black Sea Region. Economic Alternatives, 1, 106 – 116. https://www.unwe.bg/uploads/Alternatives/8_EA_1_2019_en.pdf
12. Kharaishvili, E., & Natsvlishvili, I. (2019). Markets for Georgian agri-food products and economic policy for export stimulation. Globalisation and Business, (8), 44-53.
13. Kharaishvili, E. (2017). Directions for improving the supply chain in the agro-food sector of Georgia.
14. Sagareishvili, N. (2019). The Role of State Programs to Stimulate Export of Local Agricultural Goods.
15. Georgian Farmers' Association. (2020). https://gfa.org.ge/wp-content/uploads/2020/04/GFA_COVID-19-IMPACT_GEO-FINAL_compressed-1.pdf
16. National Statistics Office of Georgia (2021). https://www.geostat.ge/en
17. Saghareishvili, M. (2019). The Impact of the Free Trade Policy on the Production of Local Agricultural Goods.
18. Liu, S. (2017). Transport and agricultural productivity: A cross-national analysis. https://www.researchgate.net/publication/314219295_Transport_and_agricultural_productivity_A_cross-national_analysis
19. United Nations Conference on Trade and Development. (2021). https://unctadstat.unctad.org/EN/Index.html
20. United Nations Conference on Trade and Development. (2021). https://unctadstat.unctad.org/wds/TableViewer/tableView.aspx?ReportId=92
21. World Export and Import of Agricultural Products (2021). https://knoema.com/cduhihd/world-exports-and-imports-of-agricultural-products
22. Koval, V., Duginets, G., Plekhanova, O., Antonov, A., & Petrova, M. (2019). On the supranational and national level of global value chain management. Entrepreneurship and Sustainability Issues, 6(4), 1922-1937. http://doi.org/10.9770/jesi.2019.6.4(27)
23. Mikhno, I., Koval, V., Shvets, G., Garmatiuk, O., & Tamošiūnienė, R. (2021). Green Economy in Sustainable Development and Improvement of Resource Efficiency. Central European Business Review, 10(1), 99-113. https://doi.org/10.18267/j.cebr.252
24. Dankeieva, O., Solomianiuk, N., Strashynska, L., Fiedotova, N., Soloviova, Y., & Koval, V. (2021). Application of Cognitive Modelling for Operation Improvement of Retail Chain Management System. TEM Journal, 10(1), 358-367. https://doi.org/10.18421/TEM101-45
25. Tsimoshynska, O., Koval, M., Kryshtal, H., Filipishyna, L., Arsawan, W. E., & Koval, V. (2021). Investing in road construction infrastructure projects under public-private partnership in the form of concession. Naukovyi Visnyk Natsionalnoho Hirnychoho Universytetu, 2, 184–192. https://doi.org/10.33271/nvngu/2021-2/184

Criteria for evaluating and selecting suppliers for maritime enterprises

Artem Lopatin[1*], *Natalia* Ishchenko[1], *Olena* Filimonova[1], and *Natalia* Rudenko[1]

[1]Petro Mohyla's Black Sea National University, 54000, Mykolaiv, Ukraine

Abstract. Supplier selection problem for maritime enterprises might be solved by using a combination of information processing methods. In this article we propose to use an integrated approach to problem analysis, specifically a multi-objectives analysis merged with analytical hierarchical process (AHP-method), k-meaning cluster analysis and Taboo Search as a way of sorting and clusterization of potential suppliers. The main purposes of this article are to provide academic researchers, maritime companies with an evaluation methodology of potential suppliers. In addition, it might be helpful for suppliers to evaluate themselves and direct their attention to the fields that matter the most for their customers. The article identifies supplier selection's criteria and sub-criteria that fits the most for maritime enterprises such as (a) parts and stores companies or (b) bunkers and lubricators suppliers 'selection accordingly. Also shows the way to gradate list of potential suppliers based on the compilation of their scores on certain sub-criteria and its weight for customer-company using analytical hierarchical process. In addition, suggested the combination of k-meaning cluster analysis with Taboo Search as a way to choose a group of suppliers for enterprises that prefer to work with couple suppliers.

1 Introduction

Picking a supplier is a difficult task for any maritime enterprise. Selection of the right supplier potentially allows a company to get the right raw materials, repairs and consumption on board, including parts, stores, bunkers and lubricators in the right quantity and at the right price at the right time. This task requires decisions based on multi-criteria qualitative and quantitative analysis of suppliers and the use of additional methods of data processing for more accurate and flexible evaluation of the supplier. In this paper we propose an option solution to this problem.

The purpose of the article is to help maritime enterprises to assess the potential of certain suppliers. For that purpose we offer a comprehensive approach to choosing a supplier. As a result of the algorithm of actions given in the article, all important qualities necessary for suppliers are selected, according to the needs of the consumer company, and the obtained data are further processed. In addition, this algorithm is interesting for suppliers because, based on it, suppliers' sales department is able to increase product sales, based on their own marketing strategy and adapt it to the results of the study. Also by this

* Corresponding author: areterik@gmail.com

study we wanted to match and adopt supplier selection criteria that are used to be used in international literature with those are met in Ukrainian science journals and presented by a Ukrainian companies.

2 Theory background and literature rewiew

Among Ukrainian scientists in substantiating the choice of supplier the greatest attention is paid to multi-criteria hierarchical analysis, these are the articles of such scientists as Bilovodskaya O.A. [1], Voznyuk N.A., Parubets E.N. etc. [2], Romashchenko O.S. [3], Kondratyuk D.M. [4] and others. The developments of foreign scientists are broader and pay attention to more diverse methods of supplier selection: Che, Wang [5], Yu, Wong [6], Khalele, Fasanghari [7], Versaci [8], Hsu [9] and others. However, comprehensive approaches to supplier selection using a combination of methods, which allows to improve existing methodologies, as well as to adapt solutions to different specialization / size / industry / capacity of enterprises, remain insufficiently covered.

Selecting the right suppliers plays a key role in any organization, as it significantly effects on the prices of proposed services and the competitiveness of the enterprise. However, in addition to the price factor in a competitive market, quality, timeliness of delivery and other factors play an important role, which complicates the choice of suppliers.

In practice, the choice of supplier by the company depends on a number of criteria, such as the proposed price, delivery time, after-sales service, flexibility in changing orders, location and financial condition of the supplier, etc. Therefore, the choice of supplier is a multi-criteria problem, which includes the assessment of both qualitative and quantitative indicators. In order for a company to choose the best supplier, it is necessary to make a compromise between these qualitative and quantitative factors.

In addition, Ghodsypour and O'Brien [10] identified two types of solutions to the problem of choosing a supplier: 1) a single supplier; 2) several suppliers.

If the chosen strategy involves a single supplier, it means that each supplier can meet all the requirements of the consumer in terms of quantity, quality and delivery of parts/stores/lubricants etc. In this case, the consumer's management needs to make only one decision - which supplier is the best. In this case, it is advisable to use a simple multi-criteria analysis and choose the supplier with the highest weighted assessment. On the other hand, if a firm plans to have several suppliers, it means that there are certain limitations in the supplier's ability to satisfy the consumer by one of the factors, or it is the costumer's principles to diversify its supplies.

3 Supplier selection methods

3.1 Analytical hierarchy process

3.1.1 Basics of analytical hierarchy process

The methodology of suppliers evaluation proposed by the authors provides for the gradual use of several methods, the combination of which will allow to select not only a specific single supplier, but also to evaluate and choose a certain combination of several suppliers. This might be achieved by using the following algorithm:

1) At the first stage, it is makes sense to use multi-criteria analysis by building a hierarchical model of suppliers.

2) Further data processing using the k-method of clustering (grouping) of suppliers for primary groups and using Taboo-search to optimize this clustering.

Multi-criteria analysis involves choosing a supplier from many conflicting factors through some compromise. Thus, multi-criteria analysis helps decision makers to evaluate a number of alternatives. However, depending on market conditions and management preferences, the criteria are of different importance and require weighing them [11].

Analytical hierarchy process (AHP) was first developed in 1971-75 in the work of Saati from 1987 [12] and was further developed in many works of both foreign [13, 14] and Ukrainian scientists [15]. This is a fairly popular method that is used due to its simplicity, no needs of specific knowledge in that field and is quite reliable for solving complex real-life problems. AHP divides the solution of a complex problem into a hierarchical system of a number of simple solutions. A matrix of pairwise comparison of these solutions is constructed, and then the vectors of each of these solutions are compared with the general vector, which takes into account the weight of each criterion in general. The procedure of AHP can be summarized in the following 4 stages [12]:

1. Construction of a hierarchical system. It must have at least 2 or more levels, and helps to relate each decision to a certain level, each of which consists of several elements. In general, AHP divides a complex problem into three levels: the overall goal; used evaluation criteria (goals); examined alternative solutions. The criteria for evaluating the effectiveness of the decision for each level should be independent of each other and go from higher priority to lower priority criteria.

2. Pairwise comparison of criteria and alternative solutions. In this manner compares the weight of the alternative solution. The result will be a ranking scale of criteria and alternative solutions for each of the criteria, taking into account their weight.

3. Calculation of local weights of hierarchy elements. Normalized values of matrix vectors, or approximate values based on the geometric mean, are used for calculation.

4. Calculation of the total weight of alternative solutions. That means that, the weight obtained for each of the criteria for each of the alternative solutions are summed. As a result of this stage, a table with ranked alternatives will be obtained. The responsible person chooses the alternative with the highest weight value.

However, in further studies [16], as a first stages, are also identified: the definition of criteria, sub-criteria and the collection of information about the supplier.

3.1.2 Review of supplier selection criteria

To select the criteria by which suppliers should be evaluated, the most frequently used criteria in Ukrainian and foreign literature were analyzed [1, 3, 9, 16, 17] and the specifics of maritime enterprises were taken into account. Thus, a hierarchical structure of criteria and sub-criteria for choosing a supplier is presented in Figure 1.

Costs are considered to be the most important criterion, as maritime enterprises operate in highly competitive markets, and for the consumer it is the final cost of proposed service (which directly depends on costs) that becomes the decisive factor. It is also one of the simplest criteria to calculate (and at the same time most transparent) criteria, as it is purely quantitative and most data can be obtained from open sources.

Quality is the most important factor for some industries in which -sized maritime businesses operate, as it is not possible to compete on price with large companies. In addition, the low quality of parts/stores/lubricants etc. determines the low quality of the proposed services, which is more likely will lead to some cost compensation, or discourage the consumer from further recourse to the maritime company, which increases marketing costs per costumer and so on.

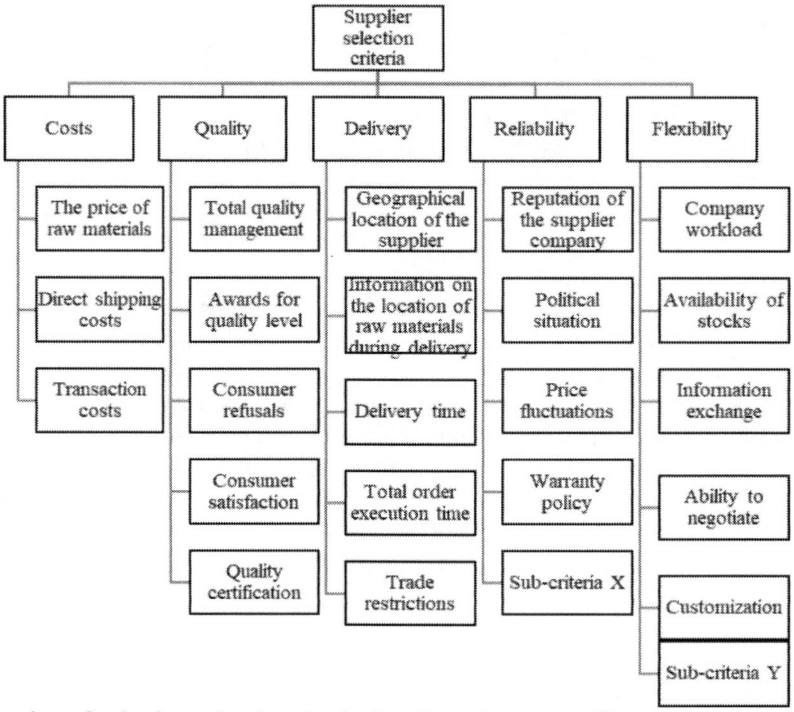

Fig. 1. Hierarchy of criteria and sub-criteria for choosing a supplier of maritime enterprise [1, 3, 9, 16, 17].

This criterion is less than directly priced because it is not quantitative, however, there is some unification of supplier evaluation by comparing quality awards and the availability of appropriate certification.

The delivery criterion strongly influences the procurement strategy, especially in sectors where the product has a short life cycle. Therefore, the total execution time of the order is crucial for the business, as it affects a number of associated costs (storage, shipment) and customer satisfaction.

The criterion of trust is somewhat biased due to incomplete information and a number of evaluative judgments, in addition, it strongly depends on the responsible person. Therefore, the proposed set of sub-criteria does not include a fairly common sub-criterion of "trust", but rather relies on indirect but open data. In addition, both internal factors of the supplier company and external national factors are taken into account. Beside, that criteria include sub-criteria X, which might differ from one type of enterprise to another. Thus, (a) quality sample test and technique problem response or (b) integrated capability for service and delivery from other suppliers suites for (a) parts and stores suppliers 'selection or (b) bunkers and lubricators suppliers 'selection accordingly.

The criterion of supplier flexibility can be considered as a tool by which the maritime company that propose service reduces the risks and consequences of an unstable environment and short-term changes in end-consumer demand. Flexibility means the ability of a supplier company to respond to short-term changes in demand and other external obstacles, as well as the ability to adapt to new conditions. This is the most difficult factor to assess, as most of the information must be found from indirect data even before the start of collaboration with new company. Moreover, the most important thing is to trust the information provided by the supplier in that case, which entails uncertainty. Yet, if company have some experience working with a number of suppliers, company can more accurately assess the alternatives by this criterion. Similarly, to the previous criteria, that criterion include sub-criteria Y, which might differ from one type of enterprise to another. Thus, (a) reachable delivery for long distance or (b) integrated capability for service and delivery from other suppliers suites for (a) parts and stores suppliers 'selection or (b) bunkers and lubricators suppliers 'selection accordingly.

In the case of the desire and ability of the maritime company to work with one parts/stores/lubricants supplier most likely it will be enough to build a hierarchical model. In this way, company can clearly assess the position of each supplier in the ranking, based on the total number of points of the integrated indicator, taking into account the personal wishes of the company's management.

In turn, if the involvement of only one supplier is not possible due to a number of related factors, it will be appropriate to further process of the data.

3.2 Data processing

3.2.1 Cluster analysis

Clustering by using k-means is one of the simplest unsupervised learning algorithms. The procedure is performed by a simple way of classifying the input data set (summarized values of evaluation points) through a certain number of pre-fixed k-clusters [26]. In Ukrainian scientific literature mostly used in a the technical field, such as energetics, computer science, fraud detection, health data etc. [19, 20, 21, 22], in economics that method is applied to the marketing task. However, there are very few articles that reveal the application of cluster analysis for the purpose of grouping and evaluation of suppliers [23], despite the fact that there are already many international sources on this field [5, 24, 25].

The basic idea is to define k centroids, one for each cluster. The placement of the centroids is a component to choose from, as it affects the end result. Many works indicate random placement of centroids, but if you place them as far apart as possible - this will reduce the number of iterations of the algorithm. The next step is to take each point (vendor evaluations) belonging to this data set and associate it with the nearest centroid. When all the points are connected, the k-number of clusters will be obtained. Next, it is needed to move the centroid to a new position and make a new binding within the resulting clusters. The resulting cycle must be repeated until the centroids k stop moving during successive iterations.

Thus, the clustering algorithm using k-means is as follows:

1. Select the points k as the initial centroid (randomly).
2. Repeat the desired number of times, based on the set coefficient k.
3. Form k-clusters by reinforcing each point to the nearest centroid.
4. List the centroid for each cluster.
5. Repeat until the centroids stop moving.

In general, this algorithm aims to minimize the objective function, in this case the error function squared (1). Target function:

$$J = \sum_{j=1}^{k} \sum_{i=1}^{n} ||x_i^{(j)} - c_j||^2 \quad (1)$$

Where, J is the objective function;

J –goal function;

k – number of clusters;

n – number of input points;

$x_i^{(j)}$ – certain point i;

c_j – centroid for cluster j.

This algorithm is sensitive to the initial randomly selected cluster centers, but cluster centers can be applied several times to reduce this effect. The main idea of the algorithm is based on an iterative search for a local minimal solution [27]. According to the author's thoughts, the number of iterations can be optimized and reduced due to the taboo search method. This method uses a strategic (general) overview and flexible memory to solve spatial problems [27], where the direction of the vector is important, which depends on its current position and history of change. The method remembers previous decisions and prevents repeating decisions that have already been determined as the worst.

3.2.2 Taboo search

The general algorithm for applying taboo search can be described by the following stages [28]:

1. Set input parameters, such as the length of the taboo list, the list of input data, the maximum number of iterations.
2. Determine the optimal starting point (in this case you can use the method of k-means)
3. According to the selected calculation function (in this case using the method of k-means) to choose local maxima of pairwise comparison, which are not specified in the tab-list.
4. In each iteration of a pairwise comparison of candidates, a taboo list with the worst values is formed and updated.
5. If one of the conditions of termination is met (the length of the taboo list is exceeded, the number of iterations is exceeded) then the optimal solution will be obtained; otherwise the current local maximum is taken as the starting point for the next iteration and the algorithm is repeated from stage 3.

Taboo search is used to solve a number of problems, such as the salesman problem, the problem of routing to return the goods, the quadratic problem of destination. Thus, according to the author's thoughts, it is promising to use taboo search in combination with the method of k-means to identify the optimal parts/stores/lubricants supplier for maritime company because by using it the searcher will not come back to the solutions, that are already 'assigned" as "not optimal" and are added to the Taboo list. Also it might affect the computer applications, that are based on Kohonen's networks (computer implementation of k-clustering) [29].

4 Conclusions

Supplier selection it is a complete task with multiple criteria must be taken into account. The article identifies compromise solutions, that according to the author's thoughts, best meet the needs of maritime companies. The proposed approach combines a number of methods to determine the optimal supplier of (a) parts and stores companies or (b) bunkers and lubricators by compilation of existing studies and extension it with sub-criteria that are used to be used in international literature with those are met in Ukrainian science journals and presented by a Ukrainian companies. Thus, in the first stage, multi-criteria analysis helps to select a specific supplier from many conflicting factors by a certain compromise among the available alternatives. In the second stage, if the manufacturer needs to have more than one supplier - the scale that is built according to the results of hierarchical analysis is processed using the method of k-means combined with taboo search. As a result, the evaluated suppliers are clustered on a certain basis, which is provided by the algorithm and respectively, the company is easier to navigate in the resulting list of companies and, if necessary, replace one supplier with another.

However, the implementation of the proposed approach, to better meet the needs of a particular enterprise, requires the participation of a responsible person / company's management to prioritize criteria and change the number of clusters depending on the specifics, location and experience with the supplier. The proposed approach is aimed mainly at maritime enterprises and can be used by them.

Further research can be direct at testing this approach at the base of particular enterprise, as well as considered in combination with other methods to solve the problem of choosing a supplier for a specific type of maritime enterprises.

References

1. Bilovodsjka, O. (2018). Ocinjuvannja postachaljnykiv u loghistychnykh (dystrybucijnykh) kanalakh komerciji innovacijnoji produkciji. Marketyngh i menedzhment innovacij, 1, 45-61.
2. Voznjuk, N., Rusynova, O., Parubec, E., Shkolenko, O., Ermyshova, S. (2019). Rozroblennja metodychnogho pidkhodu do upravlinnja postachannjam materialjnogho zabezpechennja u procesi rozvytku promyslovogho pidpryjemstva. Finansovo-kredytna dijaljnistj: problemy teoriji ta praktyky, 1(28), 313-325.
3. Romashhenko, O. (2017). Obghruntuvannja polityky vzajemodiji torghoveljnogho pidpryjemstva iz postachaljnykamy na pidpstavi kompleksnoji ocinky efektyvnosti. Ekonomika. Menedzhment. Biznes, 2(20), 75-82.
4. Kondratjuk, D. (2014). Optymaljnyj postachaljnyk jak chynnyk konkurentospromozhnosti pidpryjemstv. Ekonomika. Upravlinnja. Innovaciji. 1(11).
5. Che, Z., Wang, H. (2010). A hybrid approach for supplier cluster analysis. Computers and Mathematics with Applications, 59, 745–763.
6. Yu, C., Wong, T. (2015). An agent-based negotiation model for supplier selection of multiple products with synergy effect. Expert Systems with Applications, 42, 223–237.
7. Khaleie, S., Fasanghari, M., Tavassoli, E. (2012). Supplier selection using a novel intuitionist fuzzy clustering approach. Applied Soft Computing, 12, 1741–1754.
8. Gegovska, T., Koker, R, Cakar, T. (2020). Green Supplier Selection Using Fuzzy Multiple-Criteria Decision-Making Methods and Artificial Neural Networks, Computational Intelligence and Neuroscience, 2020, 1-26.
9. Hsu, Y.-C., Lu, H.-A., & Chu, C.-W. (2015). Evaluating and selecting maritime suppliers. Maritime Policy & Management, 43(1), 39–58.
10. Ghodsypour, O'Brien (2001). The total cost of logistics in supplier selection, under conditions of multiple sourcing, multiple criteria and capacity constraint. International Journal of Production Economics, 73(1), 15–27.
11. Dulmin, R., Mininno, V. (2003). Suplier selection using a multi-criteria decision aid method. Journal of Purchasing and Supply Management, 9(4), 177–187.
12. Saaty, R. (1987). The analytic hierarchy process: What it is and how it is used. Mathematical Modelling, 9, 3–5, 161–176.
13. Olson, D. (1996). The Analytic Hierarchy Process. Decision Aids for election Problem. Springer Series in Operations Research and Financial Engineering. https://www.springer.com/gp/book/9780387945606
14. Najib, M. (2021). Application of SWOT-AHP to develop organic food marketing strategy. Academy of Strategic Management Journal, 20(1), 1–8.
15. Kunycjka, O., Dzhyghyr, V. (2017). Vykorystannja metodu analitychnoji ijerarkhiji(metodu Saati) dlja vyznachennja miscja roztashuvannja skladsjkogho kompleksu. Naukovo-tekhnichnyj zbirnyk «Visnyk Nacionaljnogho transportnogho universytetu», 1(37), 183-188.

16. Shukla, M. (2016). Supplier evaluation and selection criteria in business performance of small and medium scale enterprise. International Research Journal of Engineering and Technology, 3(6), 70-76.
17. Ting, S., Cho, D. (2008). An integrated approach for supplier selection and purchasing decision. Supply Chain Management: An International Journal, 13/2, 116-127.
18. Tan, P., Vipin, K. (2005). Cluster Analysis: Basic Concepts and Algorithms. Introduction to Data Mining, 1, 487-568.
19. Rozen, V., Ishhuk, P., Davydenko, L. (2010). Vykorystannja metodu k-serednikh klasternogho analizu pid chas rozvjazannja zadach energhetychnoji bezpeky teoriji. Naukovo-tekhnichnoji biblioteky NU «Ljvivsjka politekhnyka».
20. Chukhraj, A., Vaghyn, E. (2011). Prymenenye metoda k-srednykh dlja formyrovanyja vneshnegho cykla obuchenyja v kompjjuternыkh proghrammakh, obuchajushhykh SQL, 2(50), 81-88.
21. Orlov, P. (2015). Suchasnyj marketyngh: analiz ta perspektyvy rozvytku. Vydannja KhNEU im. S. Kuznecja.
22. Rozhak, J., Salo, A. (2016). Proghramna systema monitorynghu anomaljnoji povedinky bankomativ pry provedenni zlovmysnykh tranzakcij. Suchasnyj stan ta perspektyvy rozvytku komp'juternykh system ta merezh. Materialy konferenciji.
23. Lotosh, O. (2019). Klasternyj analiz seghmentaciji ghaluzi. Visnyk Odesjkogho nacionaljnogho universytetu. Ekonomika, 5(78), 37-42.
24. Liu, W. Luo, T., Wang, W. (2006). Quality evaluation for three textual document clustering algorithms, J. Graduate School. Chinese Acad, 23, 640–646.
25. Kanungo, T., Mount, D., Netanyahu, N. (2002) An Efficient k-means clustering algorithm: Analysis and implementation. IEEE Trans. Pattern Anal. Mach. Intell. 24, 881–892.
26. Susana, A., Leiva-Valdebenito, F., Torres-Aviles. (2010). A Review of the Most Common Partition Algorithms in Cluster Analysis: A Comparative Study. Colombian Journal of Statistics, 33(2), 321-339.
27. Velmurugan, T. (2012). Efficiency of k-means and k-medoids algorithm for cluster Arbitrary Data Points. Int. J. Computer Technology & Application, 3(5), 1758-1764.
28. Wei, C.-T., Zuo, H., Jiang, C.-B., Li S.-F. (2017). Modeling Multilevel Supplier Selection Problem Based on Weighted-Directed Network and Its Solution, Discrete Dynamics in Nature and Society, 2017, 1-12.
29. Mahadab, A. (2020). A Study of Self- Organizing Maps(SOM) Neural Network Using Matlab. Research on Humanities and Social Sciences, 6(10), 1-22.

Indicators of the seaport ecological state of a depressed industrial hub

Irina Mayorova[1,*], *Alisa* Temerbek[2], *Oksana* Nosovckaj[3], *Valentyn* Ryabchuk[4], and *Elena* Ocheredco[5]

[1]Azov Maritime Institute National University "Odessa Maritime Academy", Department of Operation of Ship Power Plants, 19, Chornomors'ka str., 87517, Mariupol, Ukraine
[2]Azov Maritime Institute National University "Odessa Maritime Academy", Department of Navigation and Handling Ship, 19, Chornomors'ka str., 87517, Mariupol, Ukraine
[3]Azov Maritime Institute National University "Odessa Maritime Academy", Department of Natural Science and Humanitarian Disciplines, 19, Chornomors'ka str., 87517, Mariupol, Ukraine
[4]Donetsk State University of Management, 58, Karpinskogo street, 87513, Mariupol, Ukraine
[5]Azov Maritime Institute National University "Odessa Maritime Academy", Department of Management and Entrepreneurship in Maritime Transport, 19, Chornomors'ka str., 87517, Mariupol, Ukraine

Abstract. The article presents the results of the seaport ecological condition indicators study of a depressed industrial hub. The traditional ranking of indicators of ecological condition on three levels is analyzed: international, national, regional. Based on the recommendations of Eco Port PERS, the requirements of ISO 14001 and environmental management and audit, EMAS analyzes the environmental indicators that are recommended for use. The scientific novelty is the justification of the necessity to use the territorial indicator of the seaport ecological condition due to the specifics of the territorial location - first, the nature of the seaport as a city-forming enterprise - secondly, the seaport - the third, deterioration of an ecological situation of an industrial site, in - the fourth. The ecological model of the seaport is based on its division into two subsystems: production and consumer, and the link between them is transport and transport processes. The necessity of maintaining the ecological and economic balance in the seaport and on the territory of the depressed industrial hub is substantiated according to the indicators of the cargo transportation intensity, the quality of transport services and the current ecological situation of the territory as a whole, which is provided by the System of ecological condition of European ports PERS.

1 Introduction

It should be noted that the topic of "ecology of activity" (meaning any activity: production, provision of services, including transportation, loading, reloading, unloading, storage, processing, etc.) since the 70s of the twentieth century has always been one of the most important in the world, in any international forum, the problem of ecology sounded

* Corresponding author: byzp@ukr.net

particularly acute. Special attention is currently paid to the ecology of seaports. The explanation for the increased interest in the ecology of seaports is primarily due to the fact that according to historical, geographical and strategic criteria for any state, the seaport is always in the intersection center area of all transport routes and communications. An important component that explains the special attitude to seaports is due to the fact that they have always been created in industrial regions, in those places where raw materials, industrial and agricultural goods have accumulated. This has led to the fact that seaports are located in environmentally vulnerable areas; as a rule, are located within urban areas; handle large volumes of cargo that contain hydrocarbons, which inevitably affects the amount of carbon emissions and air quality. Another important circumstance that explains the increased interest in seaports, according to the authors, it is the largest volume of foreign economic activity of states is through seaports, which explains the increased level of requirements for the organization and implementation of production, economic, economic and environmental activities of seaports.

It should be added to the above mentioned that the fact that transportation through seaports is carried out using hydrocarbon fuel, as well as emissions from ships and in and their unloading, loading or reloading. Production activities and further development of seaports are directly related and interact with the air and water basin with soils, flora and fauna, as well as with the population of its territory.

2 Research methods

The theoretical basis of the work is the following fundamental provisions of the theories of transport systems, environmental safety, environmental management, transport risks, sets and the theory of environmental project management Relevance of the research. The well-known methods of differential computation are used as a mathematical apparatus.

3 Results and discussion

Apart from the above-mentioned negative environmental impact, seaports cannot properly provide environmentally friendly ship maintenance and cargo handling. Due to the active development of scientific and technological progress, the imperfection of most applied transshipment technologies, machinery, equipment and other objects of maritime transport, as well as their extremely negative impact on the environment, became especially obvious. Precisely because of the high cost of environmentally friendly cargo processing technologies, there are no opportunities for their use in production activities by Ukrainian seaports. However, as economic entities, seaports are required to comply with international and national environmental requirements.

The above circumstances, as well as issues of national security, the need to raise the image of Ukraine as a maritime power and determined the relevance of environmental indicators study of domestic seaports.

Currently, the approach to environmental indicators study of seaports is traditionally reduced to three levels, namely: international, national, regional.

The requirements of the international level of environmental safety of seaports are set out in the following documents: 17 Global Sustainable Development Goals for 2030 [1], which are supported by the UN Development Program, ratified by the Cabinet of Ministers of Ukraine in ISO 14001 [2] and Eco Port PERS [3], The International Convention for the Safety of Life at Sea SOLAS [4], the International Convention for the Prevention of Pollution from Ships MARPOL 73/78 [5], Environmental Management and Audit EMAS [6], the Strategy for the Development of Ukrainian Seaports until 2038 [7].

Among the environmental indicators by the authors of the article [8] the use of the Piguviana tax is proposed, which is calculated at the cost of bringing the ecosystem to zero. In accordance with this approach, a large number of enterprises have effective negative indicators of adverse environmental impact. It should be noted that the use of the Piguvian tax in the scientific community of both economists and ecologists is controversial. The very essence of the Piguvian tax comes down to the fact that when the marginal public interest does not coincide with the marginal interest of the enterprise owner, the latter has no incentive to reduce the value of marginal social costs; and vice versa, according to Pigou, if an enterprise provides marginal social benefit, the persons who receive it have no motivation to pay for this benefit [9]. With such a discrepancy between the ultimate interests of the owner of the enterprise and the social interests of society, the following consequences are obtained. The first consequence the party receiving social assistance does not pay for it, and the person who causes social harm does not pay for it. The second consequence is when the marginal social cost exceeds the marginal private benefit, the cost producer overproduces the product. Thus, external effects of a non-monetary nature exceed the social value (absence of environmental pollution, clean air basin, absence of soil and ground pollution), but lead to overproduction. To solve the problem of overproduction, Pigou recommended levying taxes from the violating producer, thus obliging him to pay the moral external harm he created, but at the same time it is necessary to correctly and reasonably determines the amount of this tax, i.e. it is necessary to determine the value of social costs.

Developing this idea further, it turns out that the Pigou tax controls only the scale of a particular enterprise, and not the entire number of such enterprises in the industry. If the number of polluting enterprises increases, the level of pollution and negative impact on the environment will also increase. It should also be said that it is extremely difficult, and sometimes impossible, to correctly measure social costs, since most of them are either psychological or purely individual in nature: how, for example, to measure the social benefit from the fact that the territory of a seaport is contained in cleanliness, or has well-groomed green spaces? Of course, seaports provide jobs and this effect can be measured by the number of workers and the level of their wages, but how to measure the effect when most of the seaports of Ukraine are city-forming? The way out of this situation, apparently, is to determine the specific amount of payments by enterprises, in this case, the seaport, a specific amount for a specific period of time for the use of natural resources and environmental pollution, which is currently being used.

The modern approach to the organization of environmental management in seaports is reflected in the ESPO report. The Environmental Report is an important tool of the ESPO Ecoports Network together with the Self-Diagnostic Method (SDM) and the Port Environmental Assessment System (PERS) [8]. SDM is a short checklist by which port managers can self-assess the environment. Aggregate data from SDM form the basis of annual environmental reports. Developed by the ports themselves, PERS has thoroughly established itself as a standard of environmental management for the port sector. The PERS certificate is indeed evidence of the port's voluntary compliance with environmental requirements, which it carries out independently. PERS is an EU research initiative that connects the port network, research universities and industry. More than a fifth of EU ports - EcoPorts members are PERS certified ports.

The proposed indicators of the ecological condition of the seaport in the EU include the following indicators: *A Environmental management indicators, B Environmental monitoring indicators, C Top 10 Environmental priorities, D Green services to shipping* [8]. According to the requirements of PERS, the indicators are weighed according to their significance for the environment. Significance is determined by the method of interviewing port employees and organizations working in the port, shippers' enterprises and surrounding institutions. Such certification demonstrates to a wide range of stakeholders that the seaport is competitive and meets modern ESPO requirements.

The authors consider it necessary to note that EcoPorts PERS is the only industry standard of environmental management, it is in constant open access and is becoming increasingly recognized in business circles and international organizations outside Europe.

The seaport is characterized by high compactness, density of territory development and communication saturation. The ecological and economic model of a seaport is understood as a set of interconnected economic, technical, technological, social and natural factors. The economic space of the seaport can be divided into two subsystems: production and consumption, and their connecting factor is transport, both internal transport of the seaport and external transport.

Each of these subsystems and their associated transport use environmental resources. The product of such interaction are:

-waste, which includes all types of material substances that can be disposed into the atmosphere, discharged into water (sea), buried in the ground or placed on it;

-emissions, namely emissions in the form of heat, magnetic fields, radiation, vibration.

It is impossible to consider separately the ecological situation of the seaport from the ecological situation of the city, the territory where the port is located and it is impossible not to take into account the ecological principles and approaches professed by shippers and shipping companies. The link between the seaport and the companies is transport. After all, cargo arrives at the seaport by vehicles. Mainly, road and rail transport are used for these purposes. Such an irrational by today's standards configuration of transport communications is explained by the choice of the shortest route and minimization of transport tasks, which was inherited from the socialist way of management. These principles have been used in the past when laying routes along sea coasts, beach and park areas. However, it is impossible to move the railway, which transports goods to the seaport, and no one will build bypass roads for the same purposes. Therefore, it is necessary to look for other options for a favorable solution to the existing situation.

Authors of the material [10] analyze the use of indicators of inclusive development in state (meaning country, from the word country) indicators. The main emphasis in this publication is made on the formation of the suppositions for inclusive nature management in socio-economic and environmental practices and their subsequent methodological support.

It should be noted that the concept of inclusive growth (*inclusive sustainable growth*) was nominated by the European Union Growth and Development Commission back in 2008. The concept of inclusive growth based on the "green economy" differs from the previously used strategies, provisions and principles in complex content and full involvement in government reforms, on the basis of which the diversification of national economies takes place. The European Commission included in the content of the Europe 2030 strategy the following as mandatory items: expansion of goals and growth indicators, and not just the indicator of gross national product; reducing inequality and poverty; active participation not only in the distribution of income, but also in the economic life of the country; receiving not only profits and benefits of individual ruling circles, but also noticeable improvements for all segments of the population, especially people of the "silver age", children, women; careful use of natural resources, increasing the share of renewable energy sources; widespread in all sectors of the national economy, and not only in high-tech industries with the obligatory involvement of most of the working-age population; equal opportunities for all members of society to access markets and resources; minimization of risks associated with climate change, global epidemics, massive infectious diseases and crises associated with a shortage of water resources. As it was noted above, among the environmental indicators, special attention is paid to: scarcity of water resources, careful use of natural resources, an increase in the share of renewable energy sources; an increase in the dynamics of wooded areas, the quality of the environment and the air basin, namely: the concentration of particulate matter in the atmospheric air, the intensity of carbon dioxide emissions. The solution of these tasks is possible with the creation and use of innovative ideas and products, which provides material for the development of business and scientific research, and the latter, in turn, make it possible to develop the education of the population and increase the number of jobs. Consequently, the result of introducing the concept of inclusive growth to solve environmental problems will improve the quality of life of people and their additional employment. Inclusive growth requires an accurate and thorough analysis of the environment, which can be achieved by introducing new indicators and indicators of the value of resources in projects and regional development programs.

In most scientific approaches, the criteria for the efficient operation of transport are traditionally: ensuring the efficiency of territorial logistics, maximizing profits, expanding the market segment of the business. However, nowhere is there a different approach to environmental performance indicators, which is based on a balanced, balanced approach to the organization of the transportation process, with mandatory consideration of specific priorities for resource conservation and environmental safety of vehicles and the transport process itself, which approach would take into account the existing level of pollution environment and degree of satisfaction with the quality of transport services.

Theoretically, to ensure effective regulation of the environmental situation, which is the result of the interaction of the seaport with the components of the environment to the ecological models and criteria, the following requirements are imposed:

-availability of objective and complete information on the ecological situation in the seaport, the territories surrounding the port;

- truthful identification of causal links and determination of the patterns of functioning of the system (seaport - transport - shipper) and its impact on the environment;

--development of a sufficiently complete set of scenarios for the development of the situation, which are formed on the basis of the forecast of the development of the existing environmental situation, and an integrated assessment of the results of this forecast;

-development of science-based solutions and implementation of rational regulatory measures to reduce and prevent the negative impact of the seaport system on the environment.

- constant monitoring of the state of the environment [11].

The need to maintain the ecological and economic balance in the system "seaport - transport - cargo owner (manufacturer)" is associated with a sharp deterioration of the environmental situation, which has the following features:

- increase in the volume of industrial and household waste, the size of which violates natural natural and biological processes;

- pollution of natural reservoirs and coastal sea waters with industrial and household waste;

- pollution of the environment by products of spent fuel: ash, oxides of sulfur, nitrogen, spent oil products: resins, fuel oil;

- use of materials and products containing harmful and toxic substances that are detrimental to geochemical and biological life processes;

- the presence in the environment of an increased concentration of various types of noise, vibration, radiation [13].

If we assume that the main characteristics of freight traffic in the system "seaport - transport - cargo owner" are cargo flows and the type of rolling stock that is chosen for their transportation. According to the monitoring of the state of the environment, using statistical results at a specific time t it is possible to calculate the level of marginal saturation of certain values of environmental indicators of the state of this system. The total cost of delivery of goods in the port area for a certain period of time will be presented as follows: [12]

$$\sum_{i=1}^{n} s_i = \sum_{i=1}^{n} \left(b_i \times z_i + d_i\right) \quad (1)$$

where: s_i - total cost of transportation of the i-th cargo, monetary units; b_i- volumes of cargo flows to the seaport, t; z_i - costs of industrial enterprises for the transportation of 1 ton of cargo, monetary units ; d_i - costs of the i-th enterprise for environmental measures, according to regulations and legislation, monetary units [12].

The required level of environmental safety during transportation is denoted by h_i. The current level of the current environmentally safe situation of transportation of the i-th type of products is denoted by r_i [12].

Taking the amount of fines for environmental degradation for Q_i, we obtain:

$$d_i = Q_i(h_i, r_i) = \begin{cases} Q_i(r_i), provided\ r_i < h_i \\ 0, provided\ r_i \geq h_i \end{cases} \quad (2)$$

The amount of deductions for violation of environmental requirements during the transportation of cargo is defined as the difference between the required and current levels of environmental safety. Therefore, the costs of organizing the transportation process taking into account the reimbursement of environmental contributions can be represented by expression (3) [12].

$$\sum_{i=1}^{n} s_i = \sum_{i=1}^{n} (b_i \times z_i + Q_i\,(h_i, r_i)) \quad (3)$$

In the absence of freight, no environnemental charges for violations (4) [12].

$$r_i(0, d_i) = 0 \quad (4)$$

With the growth of traffic volumes, taking into account the fact that permanent environmental deductions and fees do not change and are made in a timely manner, the amount of environmental deductions does not decrease (5) [12].

$$\frac{dr_i(b_{i,\,,}\ d_i)}{db_i} \geq 0 \quad (5)$$

At constant volume of cargo transportations and with growth of expenses for environmental protection the level of ecological deductions does not increase (6) [12].

$$\frac{dr_i(b_i, d_i)}{dd_i} \leq 0 \quad (6)$$

If freight volumes do not change, remain constant, each subsequent investment in environmental measures reduces the reduction of environmental risk (7) [12].

$$\frac{\partial r_i^2(b_i,\ d_i)}{\partial d_i^2} \geq 0 \quad (7)$$

Consider the possibility of damage from emergencies during the transportation of goods to the seaport. The possible total damage of the i-th load is denoted by F_i. The weighted average of the results of all possible emergencies in the transportation of goods to a seaport can be expressed by a mathematical expectation (8) [12]:

$$M(F) = \sum_{i=0}^{n} M(F_i) \quad (8)$$

Therefore, the ecological and economic system "seaport - transport - cargo owner (manufacturer)" must take into account the contradictions that arise in this system in a particular period of time between the needs of shippers and the constraints and capabilities of natural resources, and determine the rational relationship between consumption level, production development and environmental factors.

The share of modes of transport that participate in the transportation process can be determined by expression (9):

$$k = \frac{T_i}{\sum_{i=0}^{n} T_i} \quad (9)$$

where: T_i - number of units of vehicles of the i-th mode of transport, pieces; n - number of modes of transport, pieces [12].

The load factor of roads and access roads to the seaport can be determined by the ratio of the intensity of the i-th mode of transport to the capacity of roads (10).

$$l = \frac{G_i}{J} \quad (10)$$

where: G_i - intensity of the i-th mode of transport, pieces / hour; J - capacity of roads and access roads to the seaport, pieces / hour [12].

The following indicators can be used for the indicators "level of transport safety and ecology": the amount of emissions of harmful substances by the vehicle, the toxicity factor for the car engine.

For indicators of "quality of transport service" it is recommended to use indicators: duration of the period of delivery of freight, timeliness of arrival of the vehicle.

4 Conclusions

The research article proposes the use of ecological efficiency indicators of the ecological and economic system "seaport - transport - cargo owner (manufacturer)" which are based on a balanced approach to the organization of the transportation process, taking into account specific priorities for resource conservation and environmental safety of vehicles and the transportation process itself. This approach takes into account: the current level of environmental pollution and the degree of satisfaction with the quality of transport services.

References

1. United Nations. Department of Economic and Social Affairs. Sustainable Development (2017). Transforming our world: the 2030 Agenda for Sustainable Development. https://sdgs.un.org/2030agenda
2. International Organization for Standardization. (2015) Introduction to ISO 14001:2015. https://www.iso.org/files/live/sites/isoorg/files/store/en/PUB100371.pdf

3. EcoPorts. (2021). Port Environmental Review System (PERS): the only port sector specific environmental management standard. Developed by ports themselves. https://www.ecoports.com/pers
4. IMO. Conventions. (1974) International Convention for the Safety of Life at Sea; SOLAS. https://www.imo.org/en/About/Conventions/Pages/International-Convention-for-the-Safety-of-Life-at-Sea-(SOLAS),-1974.aspx
5. IMO. Conventions. (1978) International Convention for the Prevention of Pollution from Ships (MARPOL). https://www.imo.org
6. European Commission. (2019). Eco-Management and Audit Schem. https://ec.europa.eu/environment/emas/index_en.htm
7. Ministry of Infrastructure of Ukraine. (2020). Strategy for the development of seaports of Ukraine for the period up to 2038 Retrived from. https://mtu.gov.ua/files.pdf
8. Mikhno, I., Koval, V., Shvets, G., Garmatiuk, O., & Tamošiūnienė, R. (2021). Green Economy in Sustainable Development and Improvement of Resource Efficiency. Central European Business Review, 10(1), 99-113. https://doi.org/10.18267/j.cebr.252
9. Pigovian tax (2020). Wikimedia Foundation, Inc https://360wiki.ru/wiki/Pigovian_tax
10. Kostetska, K., Khumarova, N., Umanska, Y., Shmygol, N., & Koval, V. (2020). Institutional qualities of inclusive environmental management in sustainable economic development. Management Systems in Production Engineering, 28 (2), 15-22. https://doi.org/10.2478/mspe-2020-0003
11. EcoPortsinSights. (2020). ESPO Environmental Report –2020 https://www.espo.be/media/Environmental%20Report-WEB-FINAL.pdf
12. Gubenko, V., Nikolaenko, I. (2015). Citi Logistics: implementation of the paradigm of creative supply chains: monograph. Mariupol, PSTU, 131-135.
13. Vinogradov, A., Bogatova, J., Sinegub, I. (2014). Port ecology (Black Sea - Azov basin). Odessa. Astroprint.

Anti-corruption audit as an imperative of effective environmental management

Karina Nazarova[1], *Mariia* Nezhyva[1*], *Volodymyr* Hordopolov[1], and *Viktoriia* Mysiuk[1]

[1]Kyiv National University of Trade and Economics, Financial Analysis And Audit Department, 19, Kyoto str., 02156 Kyiv, Ukraine

Abstract. In order for the global society to emerge from the socio-economic COVID crisis, as well as to reduce its high level of eco-anxiety of various etymologies, it is necessary to restore public confidence in effective management of key segments of the national economy as an imperative of national security. In the Transparency International-19 Corruption Perceptions Index, Ukraine ranked 129th out of 180 countries (with an index of 30). The way out of the permanent state of eco-anxiety is possible only if the public information space is filled with well-founded messages about overcoming corruption risks, which are aimed at effective anti-corruption audit, which should be based on the results of basic research. The development of the region is not possible without ensuring the stable functioning of all its components: the population of the region, the state of entrepreneurship. In this case, each of the parts of the system is in correlation with each other. That is why the first direction of regional policy should be to support the introduction of eco-management in enterprises, as well as the formation of environmental behavior in the population of the region. Anti-corruption audit is an imperative for effective environmental management.

1 Background

The key problem at the global and national levels, which manifested itself globally in 2020, is the rapid growth of eco-anxiety on the planet. The COVID crisis is a manifestation of the global financial and economic, environmental, social and political crisis and is a serious challenge for the world in general and Ukraine in particular. Non-compliance with environmental obligations, conflict of interest and conflict between the provisions of different acts of equal legal force in resolving the same issue, which allows for different interpretations of norms and requirements, creates a basis for abuse. This problem can be solved by implementing institutional measures of anti-corruption audit, which will help to overcome the eco-anxiety of society and effectively combat corruption and abuse.

The fight against corruption, control of compliance with anti-corruption legislation, implementation of anti-corruption measures in public authorities, local governments and the activities of economic entities have always been the prerogative of the state. The turbulence of the economic environment and the COVID crisis are currently exacerbating threats to

* Corresponding author: marijka@ukr.net

national security and emphasizing the importance of government intervention. Conducting an effective and efficient anti-corruption audit in order to minimize the eco-anxiety of society requires justification and establishment of clear criteria for expanding the range of entities that will be subject to anti-corruption audit in Ukraine and, accordingly, determining anti-corruption audit. Implementation of anti-corruption audit in overcoming the eco-anxiety of society is impossible without the formation of methodological tools for its implementation.

The constantly growing trend to increase the amount of information that needs to be processed both within an individual economic entity and within the country as a whole, leads to the formation of innovative approaches to ensure the effective implementation of business processes. It is the dynamic change of approaches to building and conducting business, which is due to the expansion of opportunities to take advantage of the open economy, has sharply raised the issue of transition to digitalization of economic processes in the country. In turn, this form of business process support requires prompt adaptation of the country's anti-corruption system, which is designed to prevent and counter the formation of innovative corruption schemes, given the use of legally unprotected tools for digitalization of business processes. Today, the fight against corporate fraud is an important condition for economic stability of both businesses and the state. In practice, only some economic crimes are solved, while others remain in the shadows. With this in mind, more than 190 domestic companies have joined the UN Global Compact, according to which companies seek to create and maintain an ethical framework of corporate culture at both operational and strategic levels.

The term "management" is widely used in relation to various socio-economic processes occurring in the objects of economic activity in modern market conditions. Special forms of management have become widespread, focused on certain functional areas of activity of organizations and enterprises: financial management, personnel management, innovation management, quality management, etc. Environmental management is also gradually developing. Environmental management is a part of the general management system, which includes the organizational structure, planning activities, division of responsibilities, practical work, procedures, processes and resources for development, implementation, achievement of goals, evaluation of environmental policy. Environmental management is focused on regulating the interaction of society and the environment.

Increased attention from regulators to the prosecution of violators has made the fight against bribery and corruption highlighted in companies operating internationally, including Ukraine. However, according to a study by EY Global Limited, as of 2019, Ukraine ranks first in terms of corruption and fraud. In addition, in the International Corruption Perceptions Index, Ukraine ranks 130th and is rightly considered a country at high risk of corruption.

The development of the region is not possible without ensuring the stable functioning of all its components: the population of the region (its state of health, welfare, demographic situation, etc.), the state of entrepreneurship (profitability, number of jobs created, etc.). In this case, each of the parts of the system is in correlation with each other. That is why the first direction of regional policy should be to support the introduction of eco-management in enterprises, as well as the formation of environmental behavior in the population of the region. Anti-corruption audit is an imperative for effective environmental management.

2 Analysis of recent researches and publications

The issue of anti-corruption policy is the subject of research by scientists: K. Ayub [1], N. Babyak [2], L. Gnylytska [3], M. Grishchuk [4], O. Dyka [5], Yu. Zarembsky [6], A. Marchenko [7], K. Naumchuk [8], O. Parkhomenko-Kutsevil [9], K. Rostovska [10],

A. Sobakar [11], S. Timofeev [12], Y. Fidrya [13], L. Chizhevskaya [14], M. Shcherbak [15], Y. Mizik [16], A. Fadeeva [17], Yu. Slobodyanyk [18], D. Shkarovsky [19], K. Nazarova [20-23] and others. Analysis of the state and degree of scientific developments in foreign and domestic sources on the studied issues shows that corruption is not only a crime against the state and the citizen Y. Mizik [16], A. Fadeeva [17], O. Parkhomenko-Kutsevil [24], but also affects development of the system of subjects in the field of prevention and counteraction of corruption, which threatens the perception of legitimate actions of state bodies and security of citizens, negatively affects the reputation of public authorities, hinders the reform of the public administration system. At the same time, taking into account the work of P. Nieuwlands [25], J. Tillipman [26] the study will identify new trends and patterns of corruption in the system of public authorities at the local level and develop approaches to improve mechanisms to prevent and combat corruption in local government in Ukraine. Particular emphasis will be placed on the analysis of modern anti-corruption institutions as a basis for preventing and overcoming corruption, primarily in the public sector of the economy Yu. Slobodyanyk [18], B. Golovkina [27] the formation of innovative ways to develop anti-corruption institutions in Ukraine. Anti-corruption audit is relatively new to national science and practice. However, today there are a number of publications related to this issue. In the article L. Gnilitskiy [28] reveals the differences between business audit and anti-corruption audit, as well as outlines the tasks and methods of anti-corruption audit. A. Fadeeva [29] emphasize the importance of anti-corruption audit for the effective development of both individual organizations and the state as a whole. A. Novikov [30] reveals the essence of the introduction of anti-corruption compliance in the activities of domestic companies as one of the mechanisms to prevent corruption. Scientists of the Department of Financial Analysis and Audit of the Kyiv National University of Trade and Economics studied the protection of business through the sights of anti-corruption audit. Scientific research on this problem has established that the method of conducting anti-corruption audit is not fully disclosed and requires new research.

B. Bui [31] explore that accounting technologies can be used to help incentive packages achieve the dual goals of economic incentives and climate mitigation. Using the perspective of institutional logic, despite the significant financial implications of climate change, the current climate response is dominated by economic logic that justifies limited action. Further, despite the great potential, accounting technologies are not widely used in the development of incentive packages. It develops the framework for the clear integration of accounting technologies into the development, delivery and revision of incentive packages, thus ensuring economic growth and social justice while addressing the climate crisis. Yu. Vertakova [32] presents the results of the analysis of the economic, environmental and social development of the countries of the world in the modern period in conjunction with measures of state regulation, aimed at eliminating the negative impact of the development of the economic system on the environment and implementation sustainable development concept. N. Mohring [33] notes that the reduction of adverse health and environmental effects from pesticide use is currently a top priority on the agricultural policy agenda. Efficient pesticide policies must take into account farmers' application behavior, especially effects of pesticide use on economic risk. Farmers will ceteris paribus use lower quantities of herbicides, but will increase the overall toxicity of the products applied with increasing risk aversion. V. Pavlyk [34] analysis of the green investment impact on the energy efficiency gap. A. Shkuropat [35] proposes to incorporate the main indicators of "green" growth into national and regional blocks of multi-regional models, starting with the simplest options such as small econometric models of partial equilibrium, into which - based on a specially conducted analysis – the most significant factors of sustainable economic growth and exogenous parameters of public policy are included.

A special place is given to testing the effectiveness of "green" economy measures. O. Karpenko [36] introduction of the green economy model in the mechanism of state policy, development of state and regional institutions of management, transformation of social relations and modification of administrative management with saturation of its ecological dominant and development of measures on priority definition of their segments of environmentalization of national development. V. Potapenko [37] notes that green modernization of economy through analysis of barriers and drivers based on interviewing of the companies.

The purpose of the article is to determine the tasks of anti-corruption audit, features and problems of its conduct, as well as methodological directions of its conduct.

3 Materials and methods

The materials of the research were scientific works of foreign and domestic scientists on anti-corruption audit, data of the global rating of Ukraine among the countries of the world for the level of corruption, as well as statistical data on the activities of Ukrainian enterprises. The research was conducted using both general scientific methods of cognition of economic phenomena and processes, and special research methods, in particular, analysis, synthesis, systematization, theoretical generalization and comparison, statistical analysis, structural analysis, methods of expert evaluation and forecasting.

The research was conducted using the following methods: monographic – to generalize the theoretical and methodological foundations of anti-corruption audit as an imperative for effective environmental management; analysis and synthesis, induction and deduction – problem delivery, detailing and clarification of the subject of research; system analysis – to clarify the essence of the concepts of "anti-corruption audit", "imperatives", "environmental management efficiency"; graphic – for visual and schematic representation of theoretical and practical research results. The information base of the study is monographic literature, regulations of Ukraine, materials of periodicals, as well as information from the Internet.

4 Results

Corporate fraud detection studies in Ukraine indicate a trend toward growth. According to the report by PwC, World Economic Crime and Fraud Survey 2019: Survey of Ukrainian organizations, it turns out that bribery and corruption are the most common types of corporate fraud (23% of respondents). At the same time, the level of bribery and corruption in their own companies increased from 56% of respondents in 2016 to 73% in 2019. Thus, according to analysts, the level of corruption faced by business is currently growing [38]. The National Agency for Corruption's supervision of fraud in 2018 revealed 471 administrative offenses and brought them to court (Fig. 1) [39].

In the modern literature, researchers consider anti-corruption audit as a component of fraud audit, along with the facts of accounting for fraud and asset theft. Anti-corruption audit should be understood as a system of measures to manage the risks of non-compliance with the legislation of Ukraine and other countries, other normative documents, standards and ethical norms related to the fight against corruption.

Corporate anti-corruption auditing is an effective tool for combating corruption and corporate fraud. It helps to reduce the negative trends caused by the actions of company officials related to abuse of power.

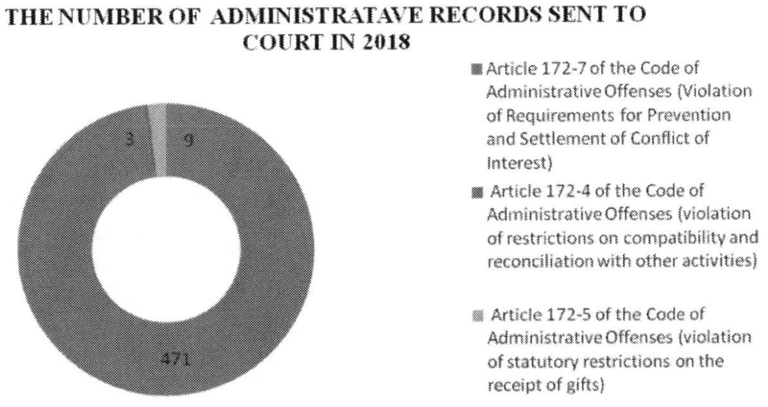

Fig. 1. Number of fraudulent administrative reports sent by the National Anti-Corruption Agency in 2018.

The peculiarity of conducting the anti-corruption audit is to evaluate the presence and observance of the company anti-corruption policy, which promotes transparency, investment attractiveness of the company and prevention of penalties.

The Law of Ukraine "On Prevention of Corruption" [40] provides for the procedure of approval of the anti-corruption program and requirements for its provisions, which are formed taking into account the assessment of corruption risks.

According to studies of modern management practices of most domestic enterprises, a serious problem faced by their management in conducting anti-corruption audit is the lack of organizational and legal support for such activities at the legislative level, random nature of existing methodological developments on the use of audit technologies, analysis of anti-corruption technologies and analysis the lack of an adaptive system for training such specialists in domestic educational institutions to the requirements of the relevant services market [41]. The main directions of anti-corruption audit are shown in Fig. 2.

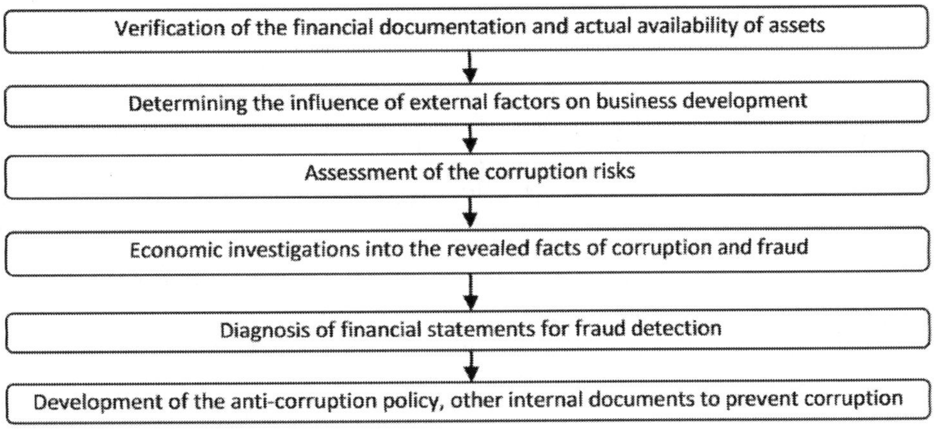

Fig. 2. Imperative directions of the anti-corruption audit.

One of the areas of anti-corruption audit is the assessment of corruption risks in the activities of a legal entity. According to the decision of the National Agency for the Prevention of Corruption [42], a legal entity must undergo an external evaluation of an audit, legal or consulting company at least once every three years.

There are many ways to misstate financial statements. The main ones include falsification of original documents, fictitious accounting records, narrowing of assets and liabilities, consolidation of heterogeneous amounts in one article reporting payments through shell companies, etc. Domestic companies may also include: fraudulent actions with company documents (forgery or falsification of documents); creation of controlled entities and receipt of operations; artificially high prices; conspiracy agreements (suppliers, customers) and false documentation, application of the "kickback" system; create accounts for outstanding works (services); the presence of atypical economic transactions and transactions at the end of the year that affect the financial results (capital, assets and liabilities) of the company; use of the company's assets for personal purposes, etc.

During the anti-corruption audit, it is advisable to take into account a possible conspiracy between the company's management and the accounting department. Owners often manipulate the company's financial results to maximize revenue, improve the company's investment attractiveness, or comply with the terms of contracts with third parties. Therefore, to identify elements of corruption, it is very important to use techniques and methods that can reveal the involvement of management in corruption crimes by distorting financial statements.

External auditors in assessing the risk of fraud can be identified by certain criteria, namely: if the company has a developed or illogical organizational structure, there is a high turnover of staff, including management staff and directors. Auditors can also warn about risky transactions carried out by the company and that they may bring significant profits in the future. Corruption schemes also include the following operations: incomplete accounting of inventory; unjustified write-off of cargo and material values for excessive losses during storage, transportation, natural disasters, etc.; disclosure of trade secrets for financial reward; payment for unprovided services or lag (services); payment of wages to employees who do not work at the enterprise, etc.

Today, an important way to fight corruption is to implement cybersecurity mechanisms. According to Allianz and WEF, cyber risks rank 2nd among all business threats and are among the top 10 risks of mankind. According to the FBI's Internet Crime Complaints Center, the damage caused by cybercrime in the world over the past five years has been $ 12 trillion. The study found that the most common types of crime faced by a company are misappropriation of property (46%), procurement fraud (33%) and personnel management (33%), and cybercrime (31%). Six thousand crimes were detected in the field of high information technology in Ukraine in 2018 [43].

The information space has become a place and an instrument of crimes, because now a crime can be committed with the help of information and communication systems. When implementing an anti-corruption audit, auditors should evaluate the experience and integrity of the company's senior management by conducting interviews, checking awareness of the company's corporate culture. During the audit, they work on the analysis of the structure of corporate business, the structure of corporate governance and corporate security, as well as study the external environment of the enterprise, the possibility of political and criminal influence. The risk map, organizational and administrative documentation, competitive environment are also analyzed.

When conducting an anti-corruption audit, it is advisable to use appropriate software, such as the software-analytical complex Global Security Assessment. An external audit of a company's corporate security eliminates or significantly minimizes the negative risks faced by most companies, which even have their own security service. It should be noted that auditors who are required to conduct anti-corruption audits must be familiar with both national and foreign legislation to prevent and combat corruption.

The specificity of environmental management is that this system is engaged in planning and regulating the conscious human impact on the environment to meet their own economic needs in terms of sustainable development of society. The separation of environmental management as an independent science has led to qualitative changes, as environmental management applies to all three named types of systems: physical, physical, biological, biological and social. Environmental management is part of the overall system of corporate governance and aims to implement the objectives of environmental policy in the field of environmental protection. Its concept is based on the model of sustainable development, which provides for a harmonious relationship between socio-economic development and the environment without harming future generations. One-sided orientation of economic activity to achieve economic results without taking into account environmental and environmental requirements cannot ensure sustainable development of society. Environmental policy of the enterprise provides: determining the activities of the organization, setting the parameters of production activities, setting environmental objectives, determining the criteria for assessing the environmental efficiency of environmental efficiency, reduction of environmental risks and reducing the cost of material resources.

The need for environmental management is determined not only by the deterioration of the eco-state, the environmental crisis, but also the natural trends of modern production, such as: differentiation of regional location of production, growth of production capacity in accordance with new technologies, the division of countries in world economy economies on those who produce toxic poisonous waste and those who concentrate it concentrate. The introduction of environmental management is also conditioned by the trends of scientific and technological progress (biotechnology, biotechnology, nuclear technology) and the emergence of environmental ecological consciousness of consciousness.

Environmental management is one of the areas of modern management and develops on its theoretical and methodological principles. It is the science of management formulates the main basic approaches to the creation of linear and functional organizations in environmental management, centralization of centralization and decentralization of management decentralization, organization of interaction and motivational processes, delegation of authority, organization of management work.

Leading environmental managers agree that by increasing the level of environmental safety of production and consumption, resource conservation and minimization of environmental risks, environmental management allows companies: to find new opportunities to save material costs, to master new environmental markets, increase the competitiveness of products; regions and countries – to improve the quality of the environment, taking into account the interests of present and future generations; nature itself – to support biodiversity and the richness of natural resources. The peculiarity of environmental management is the fact that it involves the development of environmental tasks on the basis of appropriate strategies, programs, their implementation by creating specialized structures, control over the results at all levels. First of all, this is due to the fact that environmental problems are long-term and global in nature and affect the interests of different generations. Accordingly, we propose to divide the levels of implementation and implementation of eco-management into: state, regional and local (Fig. 3). Tasks for each level of eco-management differ in their importance and implementation of functions (planning, organization, control and motivation).

The key tasks and result of environmental management today is the creation of legal, economic and social foundations for its implementation at each of the levels considered (Fig. 4).

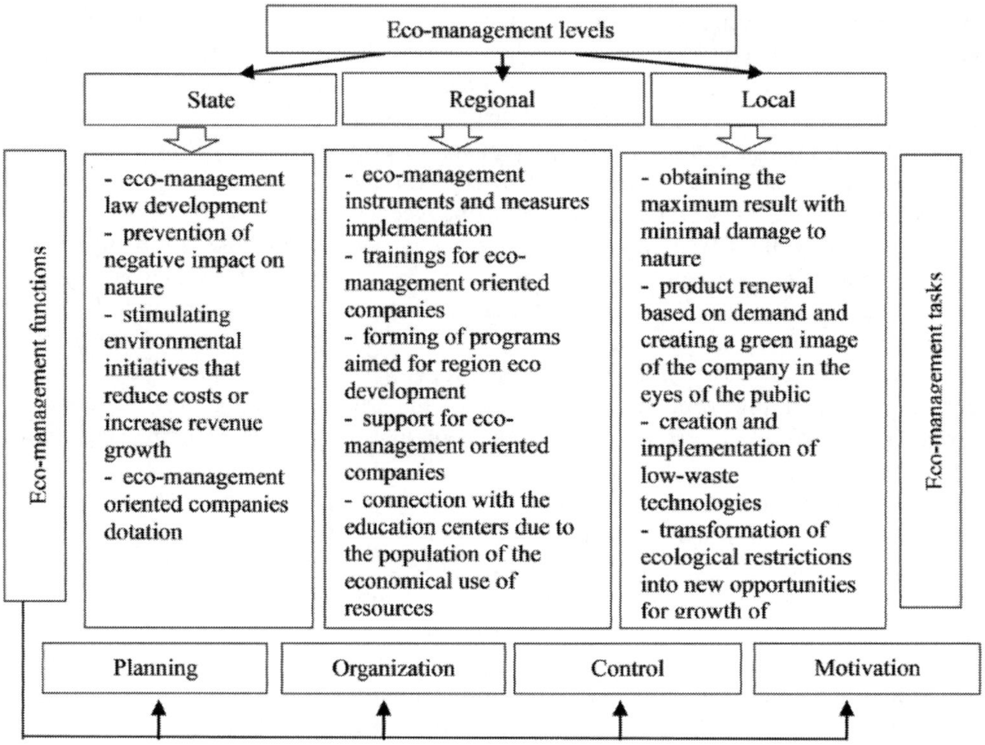

Fig. 3. Eco-management levels.

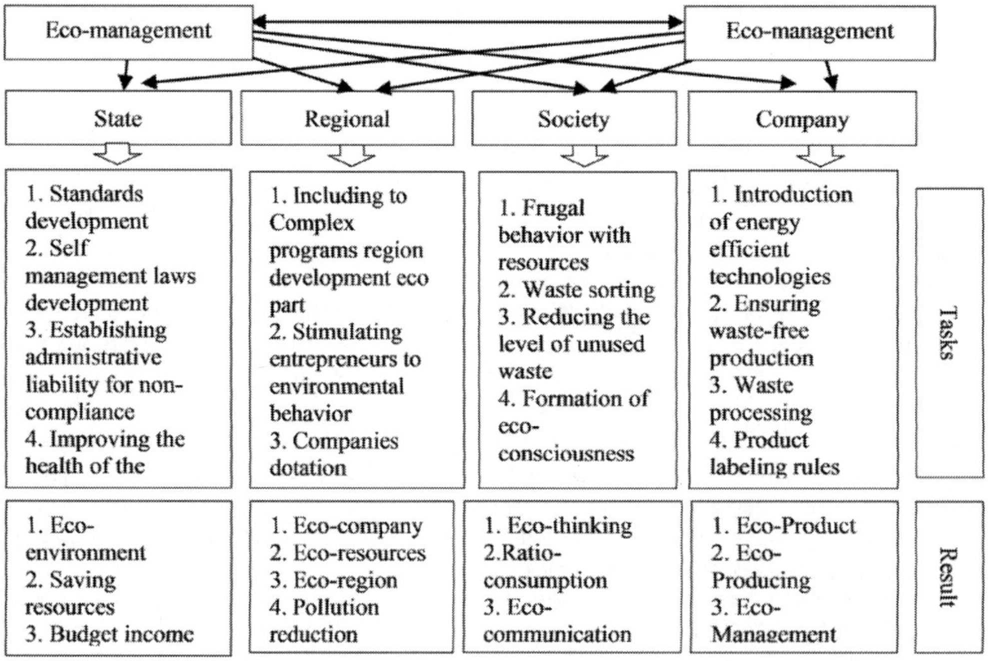

Fig. 4. Eco-management tasks.

The Environmental Efficiency Index [44] allows countries to be ranked in several categories, which are grouped into two groups: ecosystem viability and environmental health. Ukraine ranked 44th in 2016, improving its results by 25% in 10 years and found itself between Argentina – 43 and Cuba – 45 (Fig. 5).

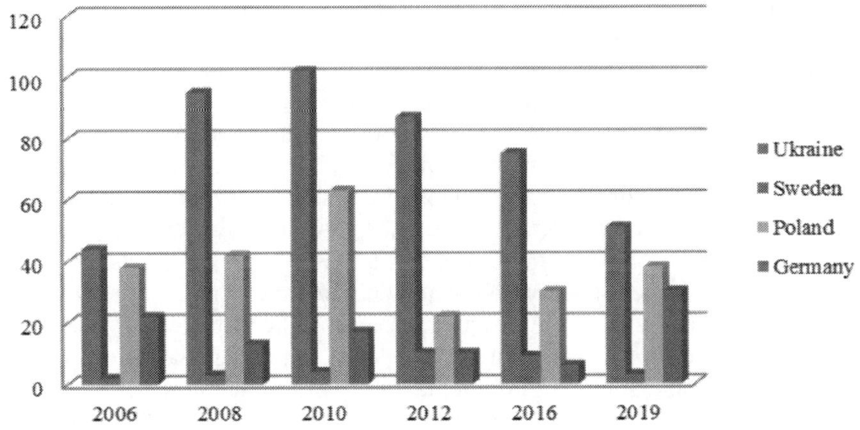

Fig. 5. Rating of Ukraine in the Environmental Efficiency Index for 2006-2019.

Enterprises need to develop the following environmental areas (Fig. 6).

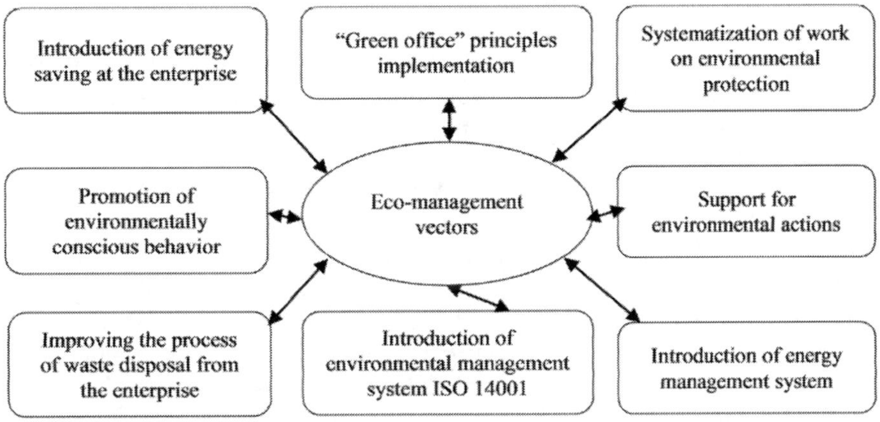

Fig. 6. Recommended eco-management vectors for companies.

Eco-management at the local level, at the enterprise level, involves the creation of a management system that should develop and implement a set of measures for rational waste-free use of resources, reducing the use of fuel and energy resources, environmentally friendly products and waste processing.

Most companies in the development of corporate social responsibility also lay down provisions on the environmental attitude to work, production, consumer, resources. Many corporations have their own recycling plants that provide reuse of raw materials. In offices, as well as in residential buildings, waste and garbage are sorted according to the possibility of their reuse or recycling. Also, each European company carries out special marking on its manufactured products in accordance with the ISO standard, which allows the consumer to determine to which type the products of this company belong.

The formation of the environmental management system takes place in a spiral: starting with the development, approval of programs and ending with the public declaration of the developed environmental policy and goals. As a basic principle, environmental policy should include the consistent improvement and prevention of negative effects on the environment. The subsequent sequence of interrelated actions in the environmental management system includes planning, organization and practical implementation, internal monitoring and control of activities carried out in accordance with the adopted policy. An obligatory part of any environmental management system is an audit (independent evaluation of the results achieved), periodic analysis and review of the management system. It is obvious that the absence or insufficient development of any stage of activity makes the whole system of environmental management inefficient.

The success of an environmental management system depends on the commitments made at all levels. Examining the level of the enterprise, we note that the system of eco-management allows the company to ensure compliance with its goals in the field of environmental protection of the region and demonstrate this to a wide range of stakeholders, which include: state environmental control; the population living in the zone of influence; consumers of products; owners; investors; insurance companies; suppliers and partners; staff of the organization; environmental community; mass media.

Environmental management is able to play an exceptional role in solving environmental problems in the exit of the regions of Ukraine from the socio-economic crisis, with a critical limitation of available financial and material resources. This conceptual vision of environmental activities, based on the ecological criteria of economic development of regions, seems promising, both at the level of methodology of economic theory and the national economic system, the global economy.

The introduction of environmental standards at the enterprise makes it possible to prevent the negative impact of economic activity on the environment, to combine direct production activities with environmental protection. The advantages that demonstrate the economic usefulness and feasibility of implementing an environmental management system and the adoption of standards in the enterprise include: increasing the competitiveness of the enterprise in domestic and foreign markets, the ability to develop new markets, including "green" light to EU markets. is even more important because of the political orientation of our country; gaining a positive image of the enterprise; training and hiring more qualified staff; improving product quality and attracting new customers, as in the minds of consumers product quality will be associated with its compliance with environmental standards; saving production costs and resources, rationalization of consumption of materials and raw materials, water, energy; reducing the risk of emergencies and the scale of their consequences; reduction of environmental payments and more effective compliance with the law, avoidance of fines and penalties by reducing emissions of harmful substances; improvement of relations with partners, investors, public authorities, which is manifested in the reduction of administrative pressure on the company by regulatory authorities through declarations, as well as the ability to obtain state support (lower interest rates on loans; lower interest rates on environmental insurance).

In order to implement eco-management at Ukrainian enterprises, it is necessary to adapt the world experience in environmental management and to implement mechanisms and tools from the already existing standards of eco-management and audit. Unfortunately, Ukrainian enterprises do not yet understand the importance of rational use of resources and protection of the environment. Moreover, only large enterprises assess the impact of their own waste and try to dispose of some, and small and medium-sized enterprises do not pay as much attention to environmental management. The reasons for this are their lack of understanding of the degree of their impact on the environment and the benefits that can be gained from the implementation of innovative environmental developments. However,

European experience shows that every European is willing to pay a higher price for an environmentally friendly product than for a regular one. Accordingly, entrepreneurs who understand this, get the maximum profits, become competitive not only in national but also in global markets, form a positive image.

Implementation of eco-management on three levels will allow the country to promote socio-economic development of regions, improve the health and life of the population, as well as enable companies to develop and implement new innovations in the field of eco-management. The implementation of environmental management will promote the socio-economic development of regions, improve the health and life of the population, as well as enable companies to develop and implement new innovations in the field of environmental management through the introduction of anti-corruption audit tools.

4 Conclusions

Internal control system testing and applying sampling method are an initial steps of anti-corruption audit doing. The approach to sampling creation should be based on auditor professional judgment, that is the main element of the anti-corruption audit as it depends on business awareness which determines potential business risk spheres.

The auditors may offer safeguards by implementing actions that minimize the causes and conditions of fraud, when they identify fraud facts. The ways to minimize the causes of fraud are to create in the company the atmosphere of social importance for each employee, material interest.

Therefore, based on the above, anti-corruption audit is an imperative for effective environmental management. The strategic goal of anti-corruption policy in the context of digitalization is to combat corruption at all levels by increasing the transparency of state bodies, respecting human and civil rights and freedoms, creating conditions for economic development, ensuring European social standards and welfare, reducing corruption in Ukraine and eliminating the causes and the conditions that determine it. Today, the digital segment of the economy is given superficial attention, which creates the preconditions for the development of corruption in this segment. The existing basic changes in the legal anti-corruption mechanism in Ukraine have laid the foundation for further improvement of the system and require access to a new level of doing business – digital.

References

1. Ayub, K. (2019). Innovative models of functioning of anti-corruption bodies in the world. Aspects of public administration, (11), 79-84.
2. Babyak, N., Bilotska, I., & Starushyk, Yu. (2015). Anticorruption compliance: basic principles and prospects for implementation in the domestic corporate sector. State and Regions. Series: Economics and Entrepreneurship, (4), 3-9.
3. Gnylytska, L. (2017). Anti-corruption audit as a means of combating corporate fraud. Independent auditor, (20), 8-14.
4. Grishchuk, M. (2019). Theoretical and legal content of the genesis of anti-corruption legislation in Ukraine. Actual problems of jurisprudence, (4), 269-275.
5. Dyka, O., & Naumchuk, K. (2019). Social aspect of corruption development in the process of development of state anti-corruption policy. Investments: practice and experience, (22), 107-111.
6. Zarembsky, Yu. (2019). Anticorruption policy of the state: administrative and legal aspect. Actual problems of jurisprudence, (4), 276-281.

7. Marchenko, A. (2020). Evolution of the system of anti-corruption bodies in Ukraine: from counteraction to prevention. Collection of scientific works of the National Academy of Public Administration under the President of Ukraine, (2), 70-79.
8. Naumchuk, K. (2020). Features of formation and implementation of state anti-corruption policy. Investments: practice and experience, (4), 134-138.
9. Parkhomenko-Kutsevil, O. (2019). Anticorruption culture as a basis for the formation of effective mechanisms for preventing corruption. Theory and practice of public administration, (2), 25-30.
10. Rostovska, K. (2018). International experience of creation and functioning of anti-corruption courts in the world. Visnyk of Kharkiv National University named after V. Karazin. Series: Right, (25), 87-89.
11. Sobakar, A. (2020). Application of anti-corruption tools in the activities of public administration entities: foreign experience and prospects of its borrowing for Ukraine. Scientific herald of the Dnipropetrovsk State University of Internal Affairs, (1), 68-74.
12. Timofeev, S. (2019). The place of anti-corruption culture in the process of preventing corruption in Ukraine. Public administration and regional development, (6), 956-978.
13. Fidrya, Y. (2017). Problems of legislative regulation of the prohibition of receiving gifts by persons covered by anti-corruption legislation. Historical and Legal Journal, (2), 135-139.
14. Chizhevskaya, L., Koretskaya, Yu., & Palamarchuk, A. (2019). Transparency of financial statements prepared in accordance with IFRS as an instrument of anti-corruption activities in Ukraine. Problems of theory and methodology of accounting, control and analysis, (1), 110-115.
15. Shcherbak, M. (2020). Legal aspect of formation and implementation of anti-corruption policy in Ukraine taking into account the experience of foreign countries. Legal Journal of Donbass, (3), 155-161.
16. Mizik, Y. (2019). Anticorruption audit. Kharkiv: KhNUMG.
17. Fadeeva, A., & Sazhina S. (2018). Anti-corruption audit as a mechanism for combating corruption. Young Scientist, (15), 41-43.
18. Slobodyanyk, Yu., Matienko-Zubenko, I., Marynich, I., & Syrotenko, N. (2020). State audit as a modern tool for preventing corruption. Financial and credit activities: problems of theory and practice, (3), 158-166.
19. Shkarovsky, D. (2019). Fighting Corporate Fraud: Tender or Necessity. Law Gazette, (11), 23-30.
20. Nazarova, K., Mysiuk, V., Gordopolov, V., Koval, V., & Danilevičienė, I. (2020). Preventional audit: implementation of SOX control to prevent fraud. Business: Theory and Practice, 21(1), 293-301. https://doi.org/10.3846/btp.2020.11647.
21. Nazarova, K., Hordopolov, V., Kopotiienko, T., Miniailo, V., Koval, V., & Diachenko, Y. (2019). Audit in the state economic security system. Management Theory and Studies for Rural Business and Infrastructure Development, 41(3), 419-430.
22. Nazarova, K., Nezhyva, M., Lositska, T., Miniailo, V., & Novikova N. (2020). Forensic-audit as an imperative of economic security and development of the company in the conditions of global transformations. Financial and credit activity: problems of theory and practice, (4), 99-106. https://doi.org/10.18371/fcaptp.v4i35.221808
23. Nazarova, K., Hordopolov, V., Zaremba, O., & Nezhyva, M. (2019). Analytical procedures of auditing within the framework of raising standards for its quality control.

Financial and credit activity: problems of theory and practice, (30), 128-135. https://doi.org/10.18371/fcaptp.v3i30.179523

24. Parkhomenko-Kutsevil, O. (2019). Formation and development of modern anti-corruption bodies of state power as a basis for preventing and overcoming corruption: theoretical and methodological analysis (Kyiv: SE "Publishing House "Personnel").
25. Nieuwlands, P. (2018). IIA Fighting Corruption: What Should Internal Auditors Do? https://ic.globaliia.org/Documents/CS-3-5.pdf.
26. Tillipman, J. (2000). The Foreign Corrupt Practices Act & Government Contractors: Compliance Trends & Collateral Consequences (GW Law Faculty Publications & Other Works. Dinah Shelton ed.).
27. Golovkina, B. (2019). Prevention of corruption. Kharkiv: Pravo.
28. Gnilitskiy, L. (2016). Anti-corruption audit in management risk corporate fraud. https://ir.kneu.edu.ua
29. Fadeeva, A. A., & Sazhin, S. S. (2018). Anti-corruption audit as a mechanism of struggle against corruption. The Young scientist, (15), 41-43.
30. Novikov, A. V. (2020). The anti-corruption compliance in the activities of Ukrainian companies as one of the mechanisms of prevention of corruption. https://www.kmu.gov.ua/en/news/uryad-priznachiv-oleksandra-novikova-golovoyu-nacionalnogo-agentstva-z-pitan-zapobigannya-korupciyi.
31. Bui, B., & Villiers, C. (2021). Recovery from Covid-19 towards a low-carbon economy: a role for accounting technologies in designing, implementing and assessing stimulus packages. American Journal of Agricultural Economics. https://doi.org/10.1111/acfi.12746.
32. Vertakova, Yu., & Plotnikov, V. (2017). Problems of sustainable development worldwide and public policies for green economy. Economic annals-XXI, (166), 4-11.
33. Mohring, N., Bozzola, M., Hirsch, S., & Finger, R. (2020). Are pesticides risk decreasing? The relevance of pesticide indicator choice in empirical analysis. Agricultural Economics. https://doi.org/10.1111/agec.12563.
34. Pavlyk, V. (2020). Assessment of green investment impact on the energy efficiency gap of the national economy. Financial markets, institutions and risks, (4), 117-123.
35. Shkuropat, A., & Stepanov, V. (2019). Methodological approaches for inclusion of factors of a "green economy" into medium term forecasting models for regional development. Ekonomichni Innovatsii, (4), 199-210.
36. Karpenko, O., Turenko, L., & Karpenko, H. (2018). Modern tools for the ensuring the eco development of the national model of the "green" economy. Ekonomichni Innovatsii, (4), 78-90.
37. Potapenko, V., Khlobystov, Y., & Kornatowski, R. (2017). Green modernization of Ukraine's economy: analysis of barriers and drivers based on interviewing of the companies. Environmental economics, (2), 50-56.
38. PwC. (2018). Fraud in Ukrainian companies. https://www.consultancy.eu
39. National Anti-Corruption Agency (2020). https://nabu.gov.ua/en/tags/national-agency-corruption-prevention
40. Law of Ukraine (2014). On prevention of corruption: the Law of Ukraine dated 14.10.2014 № 1700-VII (2014). https://zakon.rada.gov.ua/laws/show/1700-18
41. UNODC. (2021). Guidelines for the development and implementation of measures to prevent and combat corruption. https://www.unodc.org/unodc/index.html

42. Law of Ukraine (2017). The decision of the National Agency for prevention of corruption № 75 of 02.03.2017 "On approval of model anti-corruption program of legal entities". http://www.minregion.gov.ua
43. Asters (2019). Model anti-corruption program of legal entities. https://www.asterslaw.com
44. EPI (2020). Environmental Performance Index. 2020 EPI Results. https://epi.yale.edu/epi-results/2020/component/epi

Methodical approach to the choice of the variant of the shipbuilding company restructuring

Ganna Iefimova[1*], *Andrei* Labartkava[1], and *Oleksiy* Pashchenko[1]

[1]Admiral Makarov National University of Shipbuilding, 9 Geroiv Ukrainy Ave., 54007 Mykolaiv, Ukraine

Abstract. The issues of the resources use optimizing aimed at a shipbuilding company restructuring are considered, which is the main prerequisite for its effectiveness. The aim of this study is to develop a methodical approach to substantiation of the of the variant of the shipbuilding company restructuring on the basis of optimization-simulation model of its value estimation, which will take into account the algorithmically set conditions regarding the financing scheme, the type of restructuring, technological and economic relations in the system. The task of deciding on the shipbuilding company restructuring and the choice of a specific restructuring project refers to the optimization tasks of long-term planning. The limitations of the mathematical model are the company resources, which are allocated for restructuring. The main optimization criteria are maximizing net present value or minimizing the restructuring cost. The generated optimization model is designed to assess the amount of attracting for restructuring own and borrowed funds. Simulation model provides the cash flows construction. Restrictions on ensuring financial realization reflects the cash flow balance for all activities. The problem solution is achieved on the basis of an interrelated optimization, simulation models and procedures set for the formation, selection and approval of management decisions.

1 Introduction

The main task of the shipbuilding company restructuring managing in Ukraine, which are in a state of deep crisis or stagnation, is to ensure the position of the shipbuilding company in the market, at which will be overcoming the negative impact of macroeconomic factors and solving internal problems (including financial) by using all modern management capabilities, development and practical implementation of a special program of restructuring measures, which will be strategic.

* Corresponding author: hanna.yefimova@nuos.edu.ua

Company restructuring is a process of complex and interconnected change in the structure of assets, liabilities, company functions, which is initiated by a dynamic environment, based on the strategic concept of its development and aimed at improving the efficiency of its activities, expressed in increasing value [1]. The researches devoted to problems of the shipbuilding companies organizational and economic mechanism adaptation to modern conditions have found the reflection in works [2-6]. However, part of the problem of ensuring the shipbuilding restructuring programs effectiveness remains unresolved.

In our opinion, to ensure the restructuring effectiveness in modern conditions is impossible without the use in the planning process methods and tools of optimization modeling and scenario approach. The complexity of the problem is due to the need the restructuring process modeling based on a significant amount of information and taking into account many factors.

The aim of this study is to develop a methodical approach to substantiation of the of the variant of the shipbuilding company restructuring on the basis of optimization-simulation model of its value estimation, which will take into account the algorithmically set conditions regarding the financing scheme, the type of restructuring, technological and economic relations in the system.

2 Optimization model of the shipbuilding company value estimating in the restructuring process

The problem of the shipbuilding company restructuring desicion and its effectiveness estimation is one of the optimization tasks of long-term planning. If we consider it as a mathematical problem, it involves a large number of different procedures with the subsequent choice of all alternatives to the best solution for a set of parameters. The limitations of the mathematical model will be the company resources limitations allocated for restructuring, expressed by a system of equations. The main optimization criterias will be the net present value maximizing or the restructuring cost minimizing.

The first stage is the formation of restructuring alternatives, each of which is determined by the net present value, the amount of own and borrowed funds required for financing. The problem is to choose one option (among which the refusal to restructure). When there are too many alternatives, the dimension of the task increases significantly, which requires pre-ordered selection and reduction of the alternatives number.

The next step is to choose one restructuring variant. To do this, the task is formalized as follows: it is necessary to find the values of variables Z_i, $A_i(t)$, $K^+_i(t)$, $K^-_i(t)$, P_i, which provide the extremum of the selected criterion $F(PV_i, P_i, CM_i) \rightarrow extr$ under the conditions of execution of analytically and algorithmically set restrictions and conditions. Here:

$Z_i \in \{0,1\}$, $Z_i = 1$, when the i-th restructuring variant is considered, $Z_i = 0$ in other cases;

$A_i(t)$ – own funds of the restructuring company;

$K^+_i(t)$ & $K^-_i(t)$ – attraction and repayment amounts of borrowed funds in the restructuring process;

P_i – market value of the restructuring company;

CM_i – restructuring costs;

$i = \overline{0,I}$ – restructuring variant index (0 – corresponds to the variant without restructuring).

Net present value of the company after the restructuring (PV_i):

$$PV_i = \sum_{t=1}^{T} \frac{CF_t}{(1+E)^t}, \quad (1)$$

here CF_t – cash flow element, which is determined using values $A_i(t)$, $K^+_i(t)$ & $K^-_i(t)$; $t = \overline{1,T}$ – planning horizon of the company; E – discount rate.

The difference between the company income after restructuring and the income without restructuring can act as a functional $F(°)$:

$$\sum_{i=1}^{I}([PV_i + P_i - CM_i] - PV_0) \cdot Z_i \rightarrow max \quad (2)$$

Or a functional $F(°)$ may be an indicator that minimizes the restructuring cost:

$$\sum_{i=1}^{I}(CM_i - P_i) \cdot Z_i \rightarrow min \quad (3)$$

Analytically set conditions and constraints are essentially limited opportunities of company restructuring financing:

$$\sum_{i=1}^{I} A_i(t) \cdot Z_i \leq S, \quad (4)$$

here S – the maximum amount of shipbuilding company equity that can be used for restructuring measures financing.

The cost of capital is used as a discount rate, which reflects the rate of return of the shipbuilding company in the restructuring process. The discount rate is formed taking into account external and internal factors. Part of them – external – do not depend of the company and reflect the general economic conditions (inflation, general economic risk of the country, regional risk, etc.). The rest - internal - depend on the company and take into account its degree of risk (require careful analysis to avoid double counting).

The conditions and constraints of the model, given algorithmically, include:

- condition for ensuring financial realization:

$$\sum_{i=1}^{I}\sum_{t=1}^{T} CF_i(R_i(t) - C_i(t), K_i^-(t), K_i^+(t), A_i(t)) \cdot Z_i > 0, \quad (5)$$

here $(R_i(t) - C_i(t))$ – the balance of shipbuilding company flows during the restructuring;

$R_i(t)$ – company revenue during the period t;

$C_i(t)$ – company costs during the period t.

The balance of flows from operating, investing and financing activities for period t can be represented as an expression:

$$\sum_{i=1}^{I}\sum_{t=1}^{T}(R_i(t) - C_i(t) + f_1(°) - f_2(°) + A_i(t) + K_i^+(t) - K_i^-(t) - f_0(°)) \cdot Z_i (6)$$

It includes all algorithmically specified functions on the peculiarities of accounting for borrowed funds:

1) $f_0(p_n, p, D, K_i^+(t))$ – interest function for loans,

here p – interest rate on loans;

p_n – income tax rate (for companies in the general system of taxation);

D – the amount of borrowed funds raised for restructuring;

2) $f_1(p_n, f_0(°))$– capitalized interest depreciation function;

3) $f_2(p_n, \lambda, Am_i(t))$ – land tax function,

here λ – land tax rate;

$Am_i(t)$ – the residual value of capitalized interest in the period t.

4) $gD(K_i^+(t), D, K_i^-(t))$ – requirements for compliance with the conditions of attraction, repayment and servicing of borrowed funds,

here D – debt function, which is determined by the scheme of attraction, repaying and servicing borrowed funds;

gD – a set of requirements for the implementation of the restructuring financing scheme (consistency of attraction, repayment and servicing of borrowed funds, including requirements that depend on the completion date of the program, the limit conditions of the debt amount).

To take into account the residual value of the shipbuilding company, the income to be received in the post-forecast period, one of the following methods is used: Gordon's model, the method of the net assets value calculating, the method of the liquidation value calculating. The method choice depends on which scenario the restructuring and further development of the company will take place.

Given the residual value, the functionality of the optimization model will look like this:

$$\sum_{i=1}^{I}([PV_i + P_i - CM_i + RC_i] - (PV_0 + RC_0)) \cdot Z_i \rightarrow max \quad (7)$$

RC – the residual value of the shipbuilding company after the end of the planning period.

In practice the solving of the problem of choosing a shipbuilding company restructuring variant based on the formed optimization model with variables that reflect financial flows, the fact of restructuring, technological and economic links in the system, is quite difficult. The main difficulties are the presence in the model algorithmically given conditions, which cannot be described by analytical functions. The problem solution in this case can be obtained on the basis of an interconnected optimization, simulation models and procedures set for the formation, selection and coordination of management decisions.

3 Algorithm for selecting a shipbuilding company restructuring variant

Given the optimization problem complexity, it is advisable to develop an algorithm for its solution, which will present all stages of the decision-making procedure for restructuring and its effectiveness determine. The algorithm for solving the optimization problem of choosing the shipbuilding company restructuring variant on the basis of cost estimation is presented in fig. 1.

The formed optimization model is intended for an estimation of own and borrowed funds attraction amounts for restructuring carrying out. It includes a set of criteria and analytical specified conditions and restrictions. Variables are: Boolean variable Z_i; continuous variables $A_i(t)$ (the own funds amount), $K^+_i(t)$ & $K^-_i(t)$ (attraction and repayment amounts of borrowed funds in the process of restructuring). At the same time, there are restrictions on ensuring financial realization, which reflects the cash flow balance for all activities, which connects the optimization model with the simulation model.

The simulation model provides the cash flows construction. It is designed to analyze the financial realization of the restructuring variant, the formation of its financing detailed scheme (the dynamics of borrowing, repayment and servicing of borrowed funds, taking into account the availability of their own assets and reinvestment opportunities). Selection of external financing schemes carries out iteratively (at each step the scheme is specified, interest coverage and debt repayment is provided, that is the condition of financial realization).

Iterations continue until the condition of financial realization is fulfilled or the need to reject this restructuring variant is not established.

Consistency of the company structural and functional elements functioning provides emergence of effects from change, addition or removal of one «element ». The effects can be both positive and negative. To achieve positive effects in the chosen restructuring variant, it is necessary to restore possible broken chains of production or financial processes. In addition, a synergistic effect is possible in the case of the restructuring program formation, which involves the simultaneous change of several elements.

In this case, with a large number of considered variants, the optimization task dimension increases. To solve the problem of large dimension, it is proposed to use special methods that allow organizing a directed search of many variants, which can significantly reduce the number of steps in the complete search algorithms of all combinations. Algorithms based on directional search and dynamic programming method have been developed for some cases.

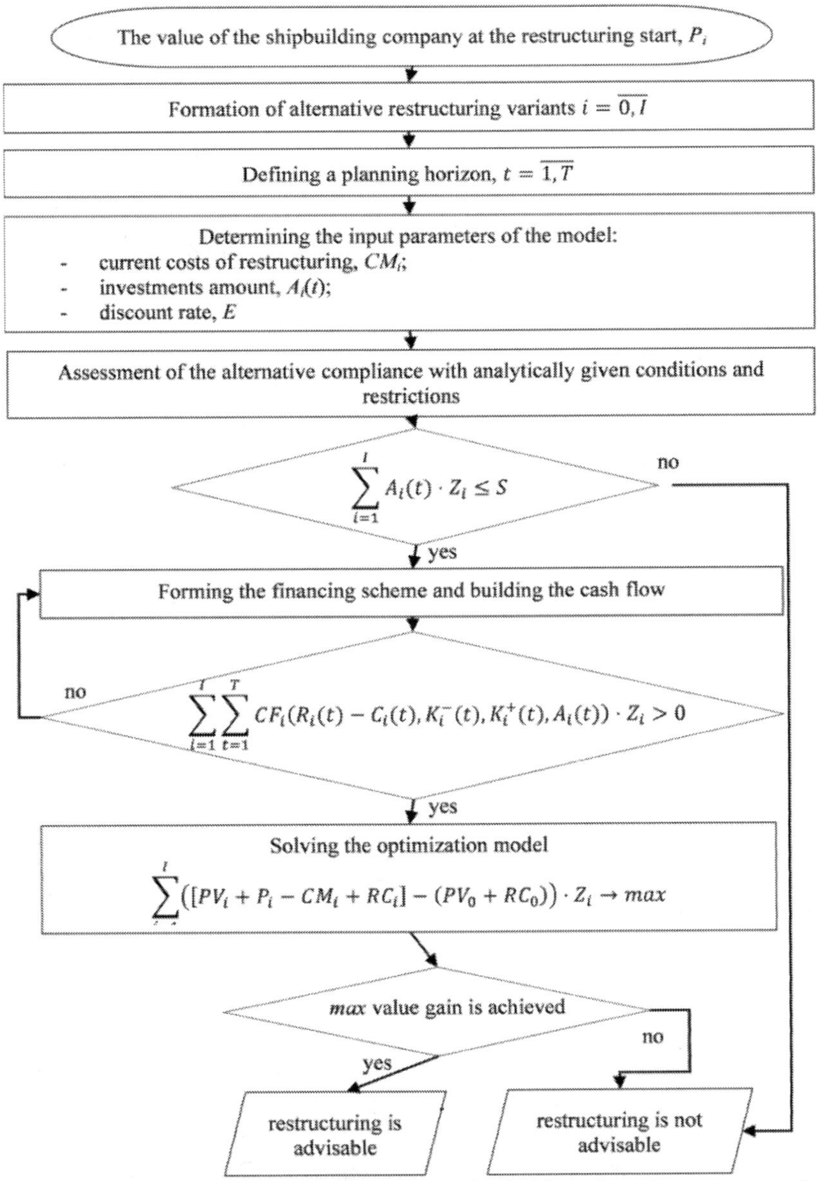

Fig. 1. Algorithm of the restructuring variant selection based on the company valuation.

4 Approaches to solving the optimization model

The task solution begins with the consideration of the maximum elements in each row of the matrix |PVi| and condition checks (5) for the corresponding elements of the coefficient matrix |Ai(t)|. If condition (5) is satisfied, then the vector of maximal elements will be the optimal solution.

The directed search algorithm in application to the task without consideration of restriction (4) consists in movement on the tree vertices received by fixing of a part of variables Zi, Zi $\in \{0,1\}$. The first level vertices are obtained by fixing in turn the elements of the system matrix first row (2), the second level vertices – fixing in turn the elements of the system matrix second row (2) etc.

For the each tree vertex we find the value of the fixed matrix element |PVi| the objective function evaluation:

$$\sum_{i<i^*} PV_i + \sum_{i\geq i^*} \overline{PV}_i \quad (8)$$

here i* – branching level; $\overline{PV}_i$ – the maximum element in the row.

Branching continues from the level top that has the maximum value of the criterion evaluation and satisfies the constraints.

The sequential search can be reduced, given the restrictions (4): from coefficients matrix |PVi| & |Ai(t)| exclude elements that cannot be included in any valid solution. For each matrix element |Ai(t)| this condition is checked:

$$\sum_{i=1}^{m-1} \overline{A_i(t)} + A_m(t) + \sum_{i=m+1}^{I} \overline{A_i(t)} \leq S, m = \overline{1, I} \quad (9)$$

here $\overline{A_i(t)}$ – the minimum element in the row.

Taking into account the limitation (5) allows the application of the described method by introducing a sequential procedure, consisting in a directional motion along the tree vertices obtained from the matrix (2), in which one of the rows is deleted corresponding to removal of a variant from group. It is expedient to compare the received admissible decisions on the qualitative characteristics corresponding to specificity of a task and to choose the optimum decision.

Dynamic programming is an approach that allows you to get the optimal solution to this problem by moving in the opposite direction – from the end of the task to the beginning [7, 8]. The large dimension problem is replaced by a series of smaller dimension problems. The basic recurrent ratio of the task dynamic programming reflects the fact that the maximum value PV, it is necessary to go from step i to the end of the task. This can be obtained by maximizing the sum of the PV values of step i and the maximum PV values that are needed to get from step i+1 to the problem end.

To apply, build a coordinate system on the plane, one axis of which reflects the restructuring elements, and the other discounted cash flows from the implementation of these restructuring program elements. Any path in the network from the initial vertex corresponds to some set of restructuring program elements. If we take the inclined arcs lengths equal to the PV value of the restructuring program variants, and the horizontal arcs lengths corresponding to the "remote" elements equal to zero, the path length connecting the initial vertex with one of the end will be equal to the total PV value corresponding to this path of the restructuring program variant. To take into account the constraint (5), the first program variant is built by drawing a horizontal arc from the origin to 1 on the elements axis that correspond to the first element removal, then build inclined arcs that correspond to the other elements development. In the second variant of the program, the horizontal arcs will correspond to the second element removal and so on. The maximum length path in each set of elements will correspond to the restructuring program variant that gives the maximum value PV, and the best version of the program is obtained by fixing the required amount of allocated funds S. The decision to choose the shipbuilding company restructuring program is made on the basis of maximizing PV.

5 The algorithm application for the shipbuilding company restructuring variant selection

Consider the task of optimization on the example of PJSC "Mykolayiv Shipyard "Ocean", where it is planned to implement a restructuring program that includes the following elements:

- production restructuring (restoration of a shipway thread and additional laying of one more, creation of additional place for placement of two more hull blocks);
- organizational restructuring (management structure change, staff reduction);

- capital restructuring (repayment of wages arrears by financing of received orders, obtaining a loan for modernization).

The main factor indicators of cash flow formation under the restructuring program of PJSC "Mykolayiv Shipyard "Ocean" are given in the table 1.

Table 1. Forecast of cash flow formation factor indicators and value of PJSC "Mykolayiv Shipyard "Ocean" for different versions of the restructuring program (thousand UAH)

Indicator	Base values (without restructuring)	Complete restructuring program	Restructuring program without element I	Restructuring program without element II	Restructuring program without element III	Restructuring program without element IV
1. Own funds of the company	156203	429703	429703	429703	429703	429703
2. The amount of own funds for restructuring	-	161513	132318	147516	161513	150116
3. The amount of borrowed funds	73541	101479	84435	79223	101479	112841
4. Discounted restructuring income	900188	1783593	1709748	1716418	1783593	1783593
5. Discounted restructuring costs	447844	476465	475502	479461	483031	476465
6. Company value	378803	1205649	1149811	1157734	1199083	1194287

The task consists in a choice of the restructuring program variant of PJSC "Mykolayiv Shipyard "Ocean" by criterion of the company capital value maximization at restriction on the general financing and taking into account restriction (4).

Compare the restructuring program variants of the PJSC "Mykolayiv Shipyard "Ocean" according to table. 1 corresponding to complete the program and its elements successively removal:

$$PV = 1205649 \text{ (thousand UAH)}$$
$$PV_{21} + PV_{31} + PV_{41} = 1149811 \text{ (thousand UAH)}$$
$$PV_{12} + PV_{13} + PV_{14} = 1157734 \text{ (thousand UAH)}$$
$$PV_{13} + PV_{23} + PV_{24} = 1199083 \text{ (thousand UAH)}$$
$$PV_{14} + PV_{24} + PV_{34} = 1194287 \text{ (thousand UAH)}$$

By limiting the financing possibility (5) and according to the company value maximizing criterion (2) we come to the conclusion about the choice of the fourth variant.

The method of dynamic programming in application to this task can be represented as a matrix method of calculation, convenient for comparison with the dichotomous programming method [9]. Indeed, both the objective function and the functions describing the problem constraints have the same network representation structures, in this case the dichotomous.

In a dichotomous representation, the task corresponding to the vertices of the network representation is the problem of optimizing the two variables function and in the discrete case is solved on the basis of a matrix representation. The first matrix corresponds to the second layer of the network. In the upper half of each cell indicates the amount of costs for different variants of the first two elements, and in the lower - the corresponding value PV. It is assumed that the variant to delete a program element is denoted by zeros in both halves of the cell. The second matrix corresponds to the third layer and so on. If there are several cells in the matrix with the same costs, then the next matrix for this cost value is taken the maximum value PV. This corresponds to the principle of Bellman's optimality for the matrix representation of the dynamic programming method. The last matrix (column) takes the optimal values of the total PV of the program variant, consisting of i-1 elements, with own funds restrictions.

Despite the fact that the general problem solution is a valid variant that optimizes the objective function value, owners are often not satisfied with one optimal variant. In most cases, it is necessary to know what optimization is achieved, and if the "losses" on other indicators are large, then also consider variants for a restructuring program, close enough to the optimal. The problem decision of a choice of the shipbuilding company restructuring program at the following stages consists in a choice of the most attractive variant from their set based on the criteria.

6 Conclusions

The optimization models use to select a restructuring variant for a shipbuilding company allows to determine the best alternative given the existing limitations. However, in its pure form, the optimization models use to decide on the restructuring is quite difficult given that not all parameters and conditions can be described by analytical functions. In this case, it is best to use optimization models in conjunction with simulations models. This will make the shipbuilding restructuring planning process dynamic, i.e. will allow to take into account algorithmically set conditions and will significantly increase the accuracy of the decision to choose the most effective option for shipbuilding restructuring.

References

1. Iefimova, Ganna V. (2014) Optimization modeling of shipbuilding enterprise restructuring efficiency. Business Inform, 3,109–113. https://www.business-inform.net/export_pdf/business-inform-2014-3_0-pages-109_113.pdf -
2. Iefimova, Ganna V., & Labartkava, Andriy V. (2015) Integration as a strategy for restructuring shipbuilding enterprises. Modern management: problems, hypotheses, researches: proceeding of material. Batumi, 45-48.
3. Iefimova, Ganna V., & Rogov, Viycheslav G. (2019) Adaptation of the shipbuilding enterprise to changes in the incentives for its development. Management and Entrepreneurship in Ukraine: Stages of Formation and Problems of Development, 1(1), 92-105. https://doi.org/10.23939/smeu2019.01.092 -
4. Burkinskyi, B. V., Laiko, O., & Losyev, M. (2018) Realization of sustainable development aims for industry and innovations on glocalization principles. Economic innovations, 20 (4 (69)), 7-22. https://doi.org/10.31520/ei.2018.20.4(69).7-22
5. Imbalances in the shipbuilding industry and assessment of policy responses. (2017) https://www.oecd.org/industry/ind/Imbalances_Shipbuilding_Industry.pdf
6. Solli-Sæther, Hans & Karlsen, Jan. (2012) Knowledge transfer in shipbuilding projects: A study of supporting mechanisms. International Journal of Project Organisation and Management, 4(3), 256-271. https://dx.doi.org/10.1504/IJPOM.2012.048224

7. Grebennik I. V., Chernaia O.S.,& Makarova E.E. (2018) Optimization of linear functions on the set of cyclic permutations with linear constraints. Control, navigation and communication systems, 3(49), 67-72 https://doi.org/10.26906/SUNZ.2018.3.067
8. Martine van den Boomen, Pieter L. van den Berg & A. Rogier M. Wolfert (2019) A dynamic programming approach for economic optimisation of lifetime-extending maintenance, renovation, and replacement of public infrastructure assets under differential inflation. Structure and Infrastructure Engineering, 15(2), 193–205 https://doi.org/10.1080/15732479.2018.1504803
9. Burkov V., Shchepkin, A., Irikov, V., & Kondratiev V. (2019) Methodology and Technology of Control Systems Development. In: Kravets A. (eds) Big Data-driven World: Legislation Issues and Control Technologies. Studies in Systems, Decision and Control, vol 181. Springer, Cham. https://doi.org/10.1007/978-3-030-01358-5_2

Evaluating performance and development priorities of port industry in Ukraine

Natalia Mihai[1*], *Larysa* Hryshyna[1], *Irina* Khmarska[1], *Elena* Pogoryelova[1], and *Nataliia* Hryshyna[1]

[1]Admiral Makarov National University of Shipbuilding, Prospekt Heroyiv Ukrayiny, 9, 54025, Mykolayiv, Ukraine

Abstract. In this publication authors have evaluated seaports performance in Ukraine as part of transport infrastructure. It is highlighted that in modern conditions development of seaports as multifunctional entities capable of performing intermodal transportation, has a strategic impact on the country and might increase its competitiveness at international level. Peculiarities of port industry development in modern economic conditions and main problematic issues are defined, among which: low congestion of port facilities; irregular mechanism of public-private partnership; loss of transit flows by seaports; low level of digitalization in the area of transport and logistics activities; insufficient level of technical, technological and environmental safety in seaports. Strategic priorities of port industry development are substantiated, considering foreign experience and modern requirements of the world community.

1 Introduction

Current stage of port industry operating in Ukraine, being a part of transport infrastructure, is characterized by insufficient development, reduced competitive position and profitability due to significant depreciation of fixed assets, as well as underdeveloped port infrastructure. Effective seaports performance depends on effective management system, implementing modern technologies and equipment capable of ensuring Ukraine's competitive advantages in the world market.

In the context of country's integration into the world community, port industry plays the key role, because seaports are the link that might accelerate economic development. Seaport – is a complex transport and logistics hub that performs various functions and provides intermodal transport. Growth of international traffic and number of participants in foreign economic activity requires special attention to performance and defining priority directions for the development of seaports in Ukraine.

[*] Corresponding author: nataliamihai@ukr.net

In recent years domestic and foreign scientists quite actively cover in their scientific papers theoretical and practical aspects of state and development of port industry in Ukraine. Among them are T.G. Logutova [1], Yu.O. Lavruschenko [2], M.V. Makarenko[3], M.M. Poltoratsky [1], A.Z. Shaikhatdinov [3], V.V.Shemayev [4]. At present, there is a large number of unsolved legal, administrative, economic problems related to forming efficient infrastructure and full performance of seaports facilities in Ukraine.

The main aim of this paper is evaluating current state and peculiarities of Ukrainian port industry performance in modern conditions and substantiation of strategic development priorities considering foreign experience.

2 Evaluating seaports performance in Ukraine

Performance of seaports in Ukraine is provided by State Enterprise "Administration of Seaports of Ukraine" established in 2013 to maintain and use port infrastructure entities of state ownership, fulfill other tasks directly assigned or given through branches, formed in each seaport [6].

State Enterprise "Administration of Seaports of Ukraine" is one of the largest state-owned enterprises established in the result of reforming port industry of Ukraine and is subordinated to the Ministry of Infrastructure of Ukraine. Enterprise structure consists of headquarters in Kyiv, main representative office in Odesa, branches "Delta-Pilot" and "Dredging fleet", and also thirteen branches in seaports of Ukraine.

Ukrainian port complex currently consists of thirteen seaports (Fig. 1), which have various specializations: cargo, passengers, finished products, food, raw materials of industrial enterprises, etc. The total cargo handling capacity of seaports is 240 million tons per year. The largest ports in Ukraine are Yuzhny, Odessa, Mykolaiv and Chornomorsk. Their main advantages are: deep-sea approaches, which allow operating large vessels; favourable territorial location. Recently concentration of cargo traffic falls upon these ports, due to annexation of Autonomous Republic of Crimea and closing of seaports located on the peninsula until restoration of Ukrainian constitutional order; significant share of private business, stevedoring companies that provide ship maintenance and loading and unloading services.

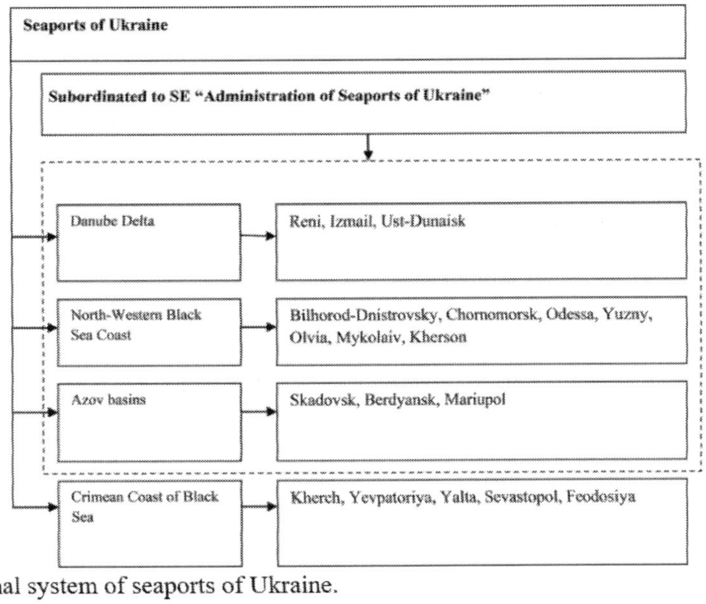

Fig. 1. National system of seaports of Ukraine.

Other seaports of Ukraine accept vessels with less draft, cargo traffic is mainly serviced by state stevedoring companies. According to State Enterprise "Administration of Seaports of Ukraine" data, 1.01 million TEU containers were loaded by Ukrainian ports in 2019 [7]. This is much higher than the world average in the container market. Table 1 shows Ukrainian seaports congestion for the period of 2015-2019 years.

Table 1. Ukrainian seaports congestion during the period of 2015-2019 years [7].

Port	Port capacity	Freight turnover, ml. tons					Fluctuations 2019 till 2015, %
		2015	2016	2017	2018	2019	
Odessa	40	24.5	25.3	24.6	24.1	25.3	+3.3
Chornomorsk	32	14.5	15.9	17.6	17.2	26.2	+80.7
Mariupol	20	14.1	7.6	13.0	6.5	6.5	-53.9
Mykolaiv	21	10.8	22.4	20.8	23.5	33.4	+209.3
Reni	14,5	1.1	0.97	1.5	0.9	1.0	-9.1
Berdyansk	1	2.3	3.8	3.2	2.4	3.8	+65.1
Bilhorod-Dnistrovsky	1	0.9	0.5	0.6	0.4	0.5	-44.4
Izmail	8,5	2.9	5.6	3.1	5.7	5.7	+96.6
Olvia	3	2,2	6.4	6.8	6.9	6.5	+195.5
Skadovsk	1	0.2	0.03	0.06	0.03	0.03	-85.0
Ust-Dunaisk	4	0.03	0.03	0.06	0.05	0.02	-33.3
Kherson	5	3.1	3.7	3.9	3.3	3.7	+19.4
Yuzhny	40	22.7	39.3	47.4	41.9	53.9	+137.4

Results of carried out analysis in table 1 indicate that only eight of the thirteen ports have positive dynamics of freight turnover. Ports of Mykolayiv (+209.3%), Olvia (+ 195.5%), Yuzhny (+ 137.4%), and Chernomorsk (+ 80.7%) show the highest growth rates of cargo handling. Decrease in cargo handling in the ports of Mariupol (-53.9.0%), Skadovsk (-85.0%), Reni (-9.1%) might be explained, on the one hand, by close location to military-political conflict in the East Ukraine, and on the other hand – the fact that these ports have problems with insufficient capacity of port railway stations and unsatisfactory railway connections. These problems should be solved at the state level. In particular, in August 2020, port of Reni received tax exemption – zero rate of land tax, which will be valid for one year and is aimed at stabilizing stevedoring company situation, promoting private stevedoring business and increasing cargo handling [9]. We also see cargo handling decrease in port of Ust-Dunaisk (-33.3%), which is explained by almost complete absence of vessel calls, imports and transit.

In this paper authors have carried out analysis of carriage of goods by sea according to cargo type (Table 2).

General tendencies of carrying goods by sea in 2015-2019 years are negative: in 2015 – 3292 thousand tons of goods were carried, and in 2019 – 2120 thousand tons, that is 35.6% less. A more detailed examination shows that during analyzed period carriage of goods increased in three positions: grain (+ 433%), coke (+ 200%), paper (+ 133.3%). At the same time carriage of specific weight of each cargo type differs in different years. Thus, if in 2015-2016 iron ore had the largest specific weight of carriage (27.1% –2015; 23.9% – 2016), various metals (35.5%; 35.5%), other cargoes (25.3%; 28.7%), then in 2017-2019 – various metals (47.2%; 53.2%; 48.6%) and grain (25.0% in 2018; 27.7% in 2019).

Table 2. Dynamics of goods carried by sea during the period of 2015-2019 years, (thsd, tons) [7, 9].

Cargo type	2015	Specific weight,%	2016	Specific weight,%	2017	Specific weight,%	2018	Specific weight,%	2019	Specific weight,%	Fluctuations 2019/2015 years, %
Total cargo volume	**3292**	**100**	**3033**	**100**	**2253**	**100**	**1892**	**100**	**2120**	**100**	-
Including											
Oil and oil products	71	2,2	53	1.74	32	1.42	30	1.6	30	1.4	-57.7
Coal	28	0.85	-	-	-	-	-	-	5	0.23	-82.2
Coke	6	0.18	27	0.89	34	1.5	23	1.2	18	0.85	**+200.0**
Iron ore	892	**27.1**	725	**23.9**	272	12.1	38	2.0	58	2.7	-94.5
Building materials	110	3.34	68	2.24	95	4.2	71	3.75	96	4.5	-12.7
Chemical & mineral fertilizers	32	0.97	17	0.56	10	0.44	12	0.63	9	0.42	-71.9
Wood	7	0.2	55	1.8	4	0.2	6	0.3	2	0.09	-71.4
Paper	3	0.09	-	-	-	-	3	0.15	7	0.33	**+133.3**
Various metals	1170	**35.5**	1077	**35.5**	1063	**47.2**	1006	**53.2**	1030	**48.6**	-11.9
Grain	110	3.34	100	3.3	204	9.05	473	**25.0**	587	**27.7**	**+433.6**
Container cargo	30	0.91	38	1.25	8	0.35	0	-	-	-	-100.0
Other cargoes	833	**25.3**	873	**28.7**	531	**23.6**	230	12.2	278	13.1	-66.6

Evaluating structure of exports and imports of main cargo groups in 2018-2020 years (Fig. 2) shows overall positive dynamics of cargo handling by stevedoring companies in seaports of Ukraine with significant predominance of exports. By specific weight, export dynamics is amounted to: 74.4% – in 2018; 76.6% – in 2019; 78.48% – in 2020.

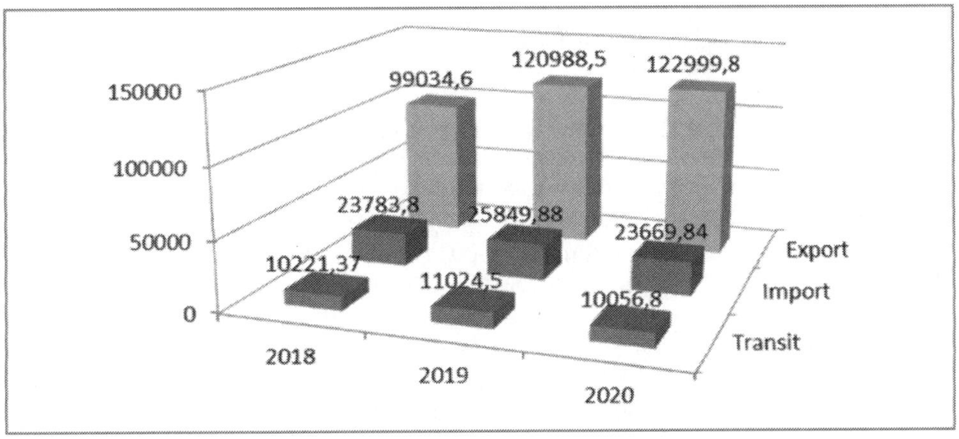

Fig.2. Dynamics of general cargo handling by stevedoring companies in seaports of Ukraine (thsd. tons).

At the same time, we observe decrease in imports and transit. Thus, if we compare volumes of transit of goods through seaports of Ukraine in 2014 and 2020, transit has decreased more than twice (-51.5%). Decrease in transit volumes may indicate irregular logistics systems and customer's distrust, cargo reorientation to other countries ports, including the Baltic States, Germany, Poland, Romania [10].

Seaports of Ukraine are called by the 15 largest shipping companies, which account for 99.5% of the total volume of containers. The top three are Maersk Line, CMA CGM, MSC, which provide more than half of entire container market in Ukraine (53.5%).

Ports of Ukraine are included in routes of the following ocean services:

1) Bosphorus Express Service (BEX) –weekly service of Ocean Alliance with calls to seaport of Odessa

2) ZIM Med Pacific (ZMP) –weekly service of ZIM company from Far Eastern ports with a call to seaport of Odessa;

3) ECUMED –weekly service of Maersk Line from Latin America with a call to the seaport Yuzhny;

4) Middle East (ME3) –weekly service of Maersk Line, which connected Ukrainian ports with the Middle East countries [11].

Unfortunately, despite the fact that indicators of Ukrainian seaports performance are generally positive, the level of competitiveness of maritime infrastructure is still low. Thus, according to the World Economic Forum (WEF) estimation, indicator of "accessibility and quality of port infrastructure" in 2018 was 3.2 out of possible seven (the estimate varies from 1 (port infrastructure is not developed) to 7 (port infrastructure meets all international standards) [12].

To sum up, analysis of seaports performance in Ukraine enabled us to define a number of problems that hinder their development and require government and business to work closely together in solving them and forming strategic development priorities for the future. Among them, in author's view the most essential are:

- low congestion of port facilities mainly due to the lack of efficient port infrastructure (low capacity of automobile and railway connections);
- irregular mechanism of public-private partnership and attraction of investments in the development of the port area;
- loss of transit flows by seaports;
- low level of digitalization in the area of transport and logistics activities;
- insufficient customs clearance procedures, corruption;
- insufficient level of technical, technological and environmental safety in seaports.

3 Priority directions of port industry development in Ukraine

Legal framework for seaports activity in Ukraine is Law of Ukraine "On Seaports of Ukraine" [6], National Transport Strategy of Ukraine until 2030, Strategy for Development of Seaports of Ukraine for the period up to 2038 [8]. In addition, Law of Ukraine "On Multimodal Transportation" is going to be approved to introduce legal mechanisms for more efficient use of multimodal freight transportation in the transport system of Ukraine. This will help to improve environmental component of transport through wider use of environmentally friendly modes of delivery, as well as optimize time and costs through implementing single agreement for multimodal transportation.

Considering that port industry in Ukraine is undergoing reforming process and ensuring level of activity that meets European standards, taking into account experience of developed countries, it is necessary to focus on Memorandum of the European Sea Ports. Priorities of European ports 2019-2024 [12].

According to the memorandum, the TOP-10 priorities for port industry development have been identified. In particular, the first priority determines that port administrations are the main component of port system, as they establish, regulate, control relationship between all stakeholders on the basis of financial transparency, competition, adaptation of port ecosystem to changes in the environment.

The second priority assumes that port authorities continue to be hybrid, that means, on the one hand, gain more financial autonomy and, on the other hand, become more socially responsible.

The third priority is that investment in European ports is an investment in European competitiveness. The main principle of trade policy should remain openness in receiving investments.

The fourth priority determines that multimodal communication is basic link between ports and logistics centers within the countries. It is stated that according to recommendations of Trans-European Transport Network (TEN-T), main seaports should have been connected by rail, road, water (where possible) by 31 December 2030. Despite the fact that such connection already exists, there is necessity for constant improving multimodal connections, in fact to invest in upgrading outdated railway and port infrastructure for their better efficiency.

The fifth priority defines the role of digitalization in improving ports operating efficiency as an essential supply chain. It is emphasized that digitalization is one of the biggest coordinators of interrelations between transport and logistics industries. And port authorities should help accelerate digitalization processes in this area.

The sixth priority highlights necessity for carrying out decarbonization measures, in particular increasing role of renewable energy in ports, which should significantly reduce carbon emissions and use of non-renewable energy resources. Responsibility for this strategy efficiency lies on International Maritime Organization (IMO), which has promised to reduce greenhouse gas emissions from vessels by at least 50% by 2050 compared to 2008, and subsequently until their complete cessation. Besides, European governments should support investments aimed at implementing strategy of port decarbonization and increasing resilience to climate change.

The seventh priority is directly related to the previous one, as it indicates that air quality – is the basis of relationship between ports and cities. In recent years, air quality has been a major environmental priority for seaport administrations in their relations with cities, as polluted air and water create significant environmental risks to public health.

The eighth priority identifies European ports tasks, which are the largest participants in the open trade environment. About 75% of European foreign trade and more than a third of its domestic trade is handled with the help of port industry, so creating any barriers to ports will negatively affect economic security of business.

The ninth priority defines that ports performance must be licensed and cannot be carried out without local community permission. 91% of ports are located near cities and are mostly perceived by population as part of urban infrastructure. Accordingly, communities demand from ports their performance transparency, dialogue with port administrations concerning security and their impact on environment. Administrations, for their part, are demonstrating initiatives to reduce external effects and improve environmental situation through independent audits and production of annual environmental reports setting a number of environmental priorities.

The tenth priority emphasizes creating pan-European policy in transport sector and its coordination with other policies at EU level (environment, customs, competition, energy, maritime affairs, research) [13].

At present, according to updated by Ministry of Infrastructure of Ukraine "Strategy for seaports development in Ukraine until 2038" we will identify the following priority areas for development of domestic port industry:

1) development of mechanisms implementing "Port Landlord" management model, according to which the state, represented by seaport administration, has the right to give investors port infrastructure entities and land areas for construction under lease or concession agreements, and investors have the right to use built terminals for a certain period of time, after which they can either extend the contract or return land together with terminal.

Action in this direction has already been taken, so in June 2020 concession agreement was signed for port of Kherson transfer to management of Risoil-Kherson LLC, which invests UAH 300 million in port and its infrastructure development. In August 2020, Ukrainian port of Olvia (Mykolaiv) was transferred to Qatari company "QTerminals" for concession. The company has committed to invest UAH 3.4 billion over the next five years in port and its infrastructure development. Projects for concession of railway and ferry complex and first and container terminals of port of Chernomorsk are under discussion with Turkish companies "CLK" (Çalık Holding) and "Busserk".

2) optimizing non-profit seaport administrations performance by their intergration on regional basis. These processes are aimed at reducing number of Administrations of Seaports of Ukraine branches in accordance with geographical distribution of ports. For example, to combine administrations of ports of Mariupol, Berdyansk, Skadovsk in Mariupol; Mykolaiv, Kherson, Olvia – in Mykolaiv; Reni, Izmail, Ust-Dunaisk– in Reni. At the same time to leave branches of administration in ports of Odessa, Yuzhny, Chernomorsk because of high ports capacity. Such optimization will significantly reduce administrative staff costs and solve the problems of ports in these regions.

3) creating mechanism for local governments participation in the processes of planning seaports development with possibility of joining seaports councils. In the conditions of decentralization processes which have been carried out in Ukraine for three years and aimed at creating capable, independent territorial communities, seaports development as a part of territorial infrastructure is able to create conditions for socio-economic development, increase job places, investments etc.

4) introduction of corporate management standards in port industry; Recently, there have been tendency in the world of port administrations corporatization. More than 50% of port administrations of European Union are corporatized commercial structures and function as independent companies Ltd-"Limited Companies", SA-"Sociétée Anonyme", GmbH, AB-"Aktiebolag", SpółkaAkcyjna, etc. [14]. Corporatization of Administration of Seaports of Ukraine will contribute to attraction of investment in port infrastructure development, simplified debt financing, reducing cost of servicing the berth line transferring this right to investors, transparency in decision-making, establishment of subsidiaries and acquisition of corporate rights.

5) development of electronic services in seaports. The main characteristic of world leading ports is digitization of all aspects of port performance, including operating, planning, design, development and maintenance of infrastructure.

Among the main innovative technologies used in leading seaports, the following should be singled out: using robotic and remotely controlled loading and unloading systems, which enable moving to container terminals automation; using Internet of Things, by implementing appropriate software for effective management of vessels traffic; implementing Blockchain technology to track cargo movement and inform stakeholders during the supply chain.

6) implementation of smart infrastructure, environmentally friendly technologies aimed at reducing harmful emissions from production processes in ports and obtaining energy

from alternative sources; The main components of smart infrastructure existing in Ukraine today, which are already used or being just implemented are: E-services (use of information and communication technologies in various sectors of economy); E-ticket – system implementation for ticket sales for any type of transport with possibility of combining transport types; Enterprise Resource Planning (ERP) – corporate information system for automation of planning, accounting, control and analysis of all major business processes, supply logistics management, personnel management, etc.; open budget – creating conditions to satisfy citizens needs getting information about using state and local budgets by Ministry of Infrastructure of Ukraine and state enterprises which are subordinated to the Ministry; open data – automatic presentation of open data registers that are filled in and maintained by state enterprises; E-transport (aviation, rail, road, water) – creating single transport model, logistics schemes at different levels, real-time optimization of freight and passenger traffic, traffic flows;

7) implementing information system "Ukrainian National Maritime Single Window" with its further integration into information systems of controlling bodies. This is information system that will enable to collect and analyze data about vessels calling Ukrainian ports in an orderly manner, giving access to country's revenue and collection authorities. The single window will allow executive branch to better interact in terms of sanitary-epidemiological, veterinary-sanitary, phyto-sanitary, ecological, radiological and other types of state control.

The single sea window will be implemented by State Enterprise "Administration of Seaports of Ukraine", and information systems of electronic interaction with it will be implemented by port community members jointly or separately. The term "port community" means two or more entities operating in a seaport; public authorities responsible for various types of control; agents, freight forwarders and entities. Ukrainian National Maritime Single Window is created on the basis of technical solutions compatible with SafeSeaNet.

8) integration of Ukrainian segment "Ukrainian National Maritime Single Window" with European Ship Monitoring System SafeSeaNet (SSN), which is aimed at exchanging between Member States all relevant information about vessels movement and exact nature of cargo carried, as well as historical information about individual vessels movement during marine environment and efficiency of maritime transport and maritime transportation.

4 Conclusions

Port industry is an important part of transport system in Ukraine. Competition in international market, a large number of problems of Ukrainian seaports are facing today, require not only implementing a set of measures, but also using all efforts by the state, private business, local communities to solve them. Country has already had some developments, highlighted in the Strategy for development of seaports in Ukraine for the period up to 2038 and a number of other regulations in this area and Ukraine is carrying out stage implementation of defined priorities.

We consider that port industry reforming must be carried out considering international experience and current requirements of world community. This will attract investments into port infrastructure modernization, introduce effective electronic services, strengthen social and environmental responsibility of port administrations and ensure state competitiveness in the world market.

References

1. Logutova, T., & Poltoratskiy, M. (2017). Analysis of the activities of Ukrainian seaports: prospects for development. Reporter of the Priazovskyi State Technical University. Section: Economic sciences, (34), 68-75.
2. Lavrushchenko, Yu. (2020). Research of world trends development of seaports. Manager. Bulletin of Donetsk State University of Management, 2(87), C. 156-166.
3. Makarenko, M., & Shayhatdinov, A. (2016) Research of directions of specialization of seaports of Ukraine. Reporter of the Priazovskyi State Technical University. Section: Economic sciences, 31 (1), 176-183.
4. Shemayev, V. (2019) Special aspects of Ukrainian sea ports authority's transition to a corporate form of governance. Development of management and entrepreneurship methods of transport, 2(67), 76-88.
5. Shemayev, V. (2017). Strategic priorities of the development of the sea ports of Ukraine. Reporter of the Kharkov National Automobile and Highway University. Problems and prospects of entrepreneurship development, 3(18.2), 9-18.
6. The Verkhovna Rada of Ukraine (2012), The Law of Ukraine "About seaports". https://zakon.rada.gov.ua/laws/show/4709-17
7. Official site of the Ukrainian Sea Ports Authority. http://www.uspa.gov.ua
8. Ministry of infrastructure of Ukraine: An updated Strategy for the development of seaports until 2038. https://www.kmu.gov.ua/news/shvaleno-onovlenu-strategiyu-rozvitku-morskih-portiv-do-2038-roku-mininfrastrukturi
9. Seaports of Ukraine and Odessa region: results of 2019. https://morhoz.odessa.gov.ua/morski-porty-ukrayiny-ta-odeskoyi-oblasti-pidsumky-2019-roku/
10. Mihai, N., (2020) Problems and prospects of development of seaports logistic infrastructure in Ukraine. Three Seas economic journal. 1(1). 53-59. http://doi.org/10.30525/2661-5150/2020-1-9
11. Krukliy, V. In 2019 seaports of Ukraine handled over 1 mn TEU containers. https://www.kmu.gov.ua/news/u-2019-morski-porti-ukrayini-perevalili-ponad-1-mln-teu-kontejneriv-vladislav-kriklij
12. World Economic Forum (2021). https://www.weforum.org/reports/the-global-competitiveness-report-2020
13. European Sea Ports Organization (2019). Priorities of European Ports for 2019-2024. Memorandum of the European Sea Ports Organization for the new Commission and European Parliament. https://www.espo.be
14. European Sea Ports Organization (2017). Trends in EU Port Governance 2016, ESPO, Brussels. https://www.espo.be/publications/trends-in-eu-ports-governance-2016

Cruise tourism in the Red Sea

Olena Budz[1], *Oksana* Romaniv[1], *Olga* Milinchuk[2], and *Dina* Yarmolyk[2*]

[1]National University of Water and Environmental Engineering, 11 Soborna St., 33028, Rivne, Ukraine
[2]Zhytomyr Polytechnic State University, 103 Chudnivska St., 10005, Zhytomyr, Ukraine

Abstract. Cruising is a lifestyle that involves living for extended time on a boat while traveling from place to place for pleasure. Cruise lines have a dual character; they are partly in the transportation business, and partly in the leisure entertainment business, a duality that carries down into the ships themselves, which have both a crew headed by the ship's captain, and a hospitality staff headed by the equivalent of a hotel manager. The business is extremely volatile. Cruise lines frequently sell, renovate, rename their ships just to keep up with travel trends. That is why the aims of the study were to focus on the development of cruise tourism in the Red Sea, to characterize the major cruise lines in the Red Sea. There are 11 cruise lines that operate in the Red Sea. Cruise line rating scores were developed in this study. The Royal Caribbean International is a leader. The Oceania Cruises is an outsider. The primary reasons that consumers will be motivated to book a cruise: good value offered by the cruise lines; love of cruising; get away with friends or alone; family reunions, honeymoons/weddings/anniversaries; try out new sustainable ships; shorter cruises; and «work and play» opportunity.

1 Introduction

Travel on cruise ships may be referred to as cruising. Cruising generally refers to trips of a few days or more, and can extend to round-the-world voyages.

A cruise line is a company that operates cruise ships. Among cruise lines, some are direct descendants of the traditional passenger shipping lines, while others were founded from the 1960s on specifically for cruising. The business is extremely volatile; the ships are massive capital expenditures with very high operating costs, and a slight dip in bookings can easily put a company out of business. Cruise lines frequently sell, renovate, or simply rename their ships just to keep up with travel trends. A wave of failures and consolidation in the 1990s has led to many lines existing only as "brands" within larger corporations, much as a single automobile company produces several makes of cars. Brands exist partly because of repeat customer loyalty, and also to offer different levels of quality and service. For instance, Carnival Corporation owns both Carnival Cruise Lines, whose former image were vessels that had a reputation as "party ships" for younger travellers, but have become large, modern, and extremely elegant, yet still profitable, and Holland America Line, whose ships cultivate an image of classic elegance.

* Corresponding author: dina.yarmolyk@gmail.com

Currently the five largest cruise line operators in the world are Carnival Corporation & plc, Royal Caribbean Cruises Ltd., Star Cruises, Mediterranean Shipping Company (MSC) Cruises and Louis Cruises Lines [1].

Most cruise ships are centered around the Caribbean and Mediterranean.

The Red Sea is one of the world's top mass tourism destinations and tourist flows to this region are constantly increasing. However, the development of cruising in this region not described in scientific publications enough. It is therefore very important to explore the cruising in the Red Sea.

2 Characteristics of the major cruise lines in the Red Sea

There are 11 cruise lines operate in the Red Sea: Azamara Club Cruises, Costa Cruises, Cunard Line, Holland America Line, MSC Cruises, Oceania Cruises, Princess Cruises, Regent Seven Seas Cruises, Royal Caribbean International, Seabourn Cruise Line, Silversea Cruises.

Cruise line rating scores were developed for specific characteristics in this study: 0 point – no response; 1 point – is not on all ships; 2 points – is available on all ships.

It is best to focus on the services, programs when making evaluation rather than using the price, client's lifestyle or client's age as deciding factors. The following characteristics have been included for compiling the rating (table 1).

The Royal Caribbean International is a leader (144 points). The Oceania Cruises is an outsider (95 points) (Fig. 1).

Let's describe the key characteristics of the leader and outsider of our rating now.

Royal Caribbean International is a leading, innovative cruise vacation company. Guests are active travelers who are looking for new experiences, a wide array of choices for the whole family (from activities onboard all ships like rock climbing, surfing and ice skating to shoreside adventures like glacier trekking and exploring Mayan ruins). The brand is distinguished by interesting Gold Anchor Service, shoreside adventures, innovative ships.

Oceania Cruises is the world's upper-premium cruise line. This includes the finest cuisine at sea, carefully crafted by world-renowned Master Chef Jacques Pepin. Each ship offers the flexibility of open-seating dining in four unique restaurants and all specialty restaurants are complimentary. The warm and welcoming country club atmosphere encourages guests to dress comfortably, suits, tuxedos and gowns are never required. Oceania Cruises sails to more than 350 ports around the globe. Featuring a fleet of intimate and elegant mid-sized ships, Oceania Cruises is focused cruising overnight and multi-day stays in the world's most alluring ports. Therefore, the last described cruise line was lower in the rating due to fewer services. But it is also a prestigious cruise line.

Now thoroughly characterize the tourism product of one of the cruise lines.

For a cruise holiday of truly biblical proportions, there can be no better destination than the Red Sea – on board one of Mediterranean Shipping Company (MSC) Cruises' luxury liners.

Best of all, tourists can enjoy their Southern Sea experience from the first-class comfort of one of MSC Cruises' stylish cruise liners [2-3].

MSC Cruises has introduced what it calls the "perfect comprehensive itinerary" combining the cultures and nature experiences offered by Red Sea destinations. In addition to a full range of shore excursions, including historical tours in Egypt, Jordan and Israel, the cruises feature relaxation and snorkelling opportunities, and a variety of desert tours.

Themed menus and entertainment onboard are also designed to bring the region to life. According to MSC Cruises, menu items range from Mediterranean dishes to Arabian appetizers and traditional main courses, including marinated meats and fish dishes, tagines and couscous.

Table 1. Characteristics for making evaluation of cruise line.

Factors	Characteristics	Factors	Characteristics
1. Stateroom Amenities	- TV/Radio - CD Player - In-Stateroom Movies - iPod Docks/Stereos - In-Stateroom Internet Access - Direct Dial Telephone - Private Safe - Minibar/Refrigerator - Hair Dryer	5. Special Programs/ Services	- Weddings/Honeymoon Packages - Guest Lecturers/Educational Programs - Cell Phone Service - iPod and/or iPhone Rental - Special Dietary Needs (Advance Notification Required) - Wheelchair Accessibility
2. Dining	- Traditional Assigned Seating - Open Seating - Formal Nights - Specialty/a la carte Venues - Coffeehouse/Patisserie - 24 Hour Food Service Available	6. Sports and Fitness	- Cardio Equipment - Weights (free weights and/or machines) - Personal Trainer - Total Fitness Program - Group Fitness Classes - Yoga/Pilates - Jogging Track - Basketball - Volleyball - Multi-use Sports Court - Paddle Tennis/Ping Pong - Rock Climbing Wall - Bowling Alley - Miniature Golf - Watersports Platform - Snorkel Lessons - Discover SCUBA /Resort Course - SCUBA Certification - Golf Programs - Golf Driving or Simulator - Clinics/Lessons - Onboard Pro - Golf Packages
3. Children and Families	- Family Staterooms - Reduced Cruise Rates - Babysitting Available - Cribs Available - Organized Age-Specific Programs - Games/Contests - Organized Parties and Events - Organized Sporting Activities - Cartoons - Movies - Educational Programs - Environmental Awareness Programs - Children-Only Pool - Children's Playroom - Teen Center or Disco - Teen Counselors - Youth Counselors	7. Spa/ Wellness	- Aromatherapy - Body Wraps - Mud Therapy - Ionithermie Treatments - Reflexology - Thalassotherapy - Massage - Facials - Manicure/Pedicure - Full Service Beauty Salon - Teeth Whitening - Nutritionist - Fitness Evaluations - Steam Room or Sauna - Jacuzzi or Whirlpool
4. Meetings and Conferences	- Meeting Coordinator - Meeting Space - Conference Dining - Private Party/Events - Business Center Services - Audio Visual Equipment - Computer Equipment - Email Access		

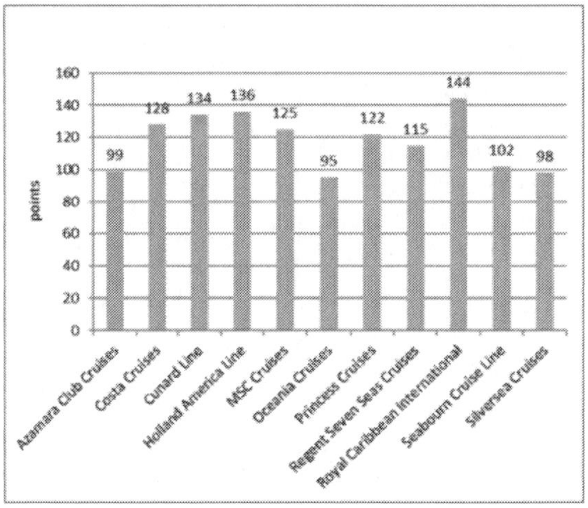

Fig. 1. Rating score of cruise lines (evaluated in points).

The ship's theatre will show the movies Cleopatra and A Queen's Life, "taking guests back to ancient Egypt". And for relaxation are sandy beaches and crystal clear water – "a haven for sunbathing or setting off on a discovery of the Red Sea's underwater world," according to MSC Cruises. Related tours are snorkeling excursions, trips in glass-bottomed boats, camel rides or jeep drives in the desert, or just floating in the saline waters of the Dead Sea for its alleged therapeutic effect [3].

Consider one typical cruise. [4] Parameters of studied route:

- Departing from: Sharm Ash Sheikh.
- Ports of call: Sharm Ash Sheikh, Safaga, Elat (Eilath), Aqaba, Sokhna Port.
- Duration: 7 nights.

Prices cruises in the Red Sea are lower than in other destinations [5].

3 Market segmentation

The cruise industry is segmented into types of cruises based on certain characteristics that define the on-board lifestyle and overall vacation experience.

Cruise lines fall also into loose categories that are generally termed Mainstream, Premium, and Luxury [6]. The First-Class cruise lines, (also known as the mass market lines), and the Premium cruise lines are resort-style vacations offering a broad base of services, destinations and amenities that are affordable to most travelers.

Over the past few years these lines have mimicked one another's successes so much that the two segments have merged in their "mass-clusivity" to create a large contemporary segment of the cruise market. Such cruise lines predominate in the Red Sea.

First Class Cruise Lines: Costa Cruises, Mediterranean Shipping Company (MSC) Cruises, Royal Caribbean International. Key Word: Quantity

Premium Cruise Lines: Azamara Club Cruises, Holland America Line, Oceania Cruises, Princess Cruises. Key Word: Quality.

The Figure 2 below illustrates a number of variables that predict cruise behaviour, and how they might aggregate to form market segments.

Consumer marketing segmentation was carried out in %, using the methodology of scientific works [7-12] (table 2). «Explorers» are the largest segment among cruising consumers in the Red Sea.

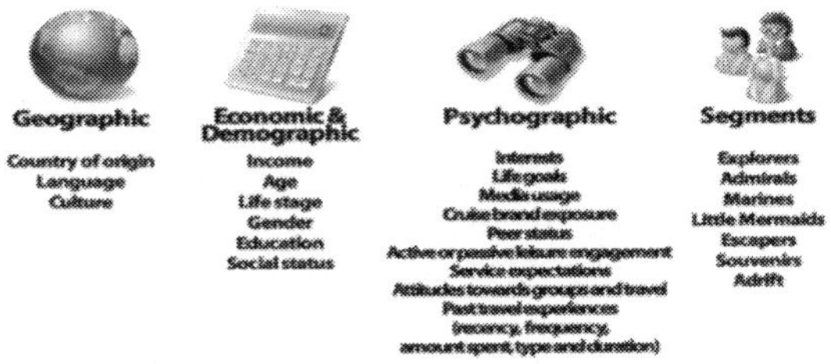

Fig. 2. A number of variables that predict cruise behaviour [13].

Table 2. Customer cruise segmentation to all destinations of Red Sea [13].

Segment	Estimated % of total market by trips to all destinations of Red Sea
Explorers – These people take four or more vacations per year, have disposable incomes and take longer cruises, exotic cruises and cultural learning cruises. Education and social causes are important to them. So are making friends and socializing. It's a more saturated segment, but one that is lucrative and important to satisfy to retain their business. This group also represents future opportunity, as more couples become empty nesters and retired upscale boomers.	25
Admirals – These people have selected their preferred cruise provider and seek a traditional experience. They tend to ritualize their travel experience and don't usually experiment unless their favorites start to become stale or so radically different the attributes they admired become unrecognizable. Great cruise consumers, they tend to be older and a good, loyal customer base but offer less opportunity for growth.	20
Marines – This desirable yet elusive segment is made up of upscale, motivated and active young professionals. They are most likely to snorkel, para-sail, surf and rock climb. Whether new or experienced cruisers, they are always auditioning better ships. They are intellectually curious, media-involved, and they perceive value in not only the appearance of being active but also the reality of learning and being challenged. Cruise companies can grow well in this segment. They are the logical target for active ship design strategies as well as expanding Internet marketing.	15
Little Mermaids – This segment is made up of upper middle class families. They are experiencing an increase the pace of daily activity and a crunch for time. With every non-working moment devoted to family errands (stopping at the Home Depot to pick up an attachment for the air pump for the kids pool or running to Target for a new basketball for the son's friends birthday party) they are looking to maximize leisure activity as a family experience that includes opportunities for real quality-bonding.	13
Escapers – This is a desirable segment and probably the core of the cruise market. They are just looking to get away. All-inclusive. No complications, no worries. From their point of view, after having spent a hectic year, they have earned the pleasures of doing nothing but sitting by the pool, seeing a few sites and relaxing. They are somewhat price sensitive but will always find the money for the trip they deserve.	12
Souvenirs – These people have jobs (not careers) and lives (not lifestyles). Lacking intense interest in the world outside they are primarily focused the internalized experience of the moment. They tend to take a cruise vacation only when there's a "really good deal" that everyone's talking about.	8
Adrift – There is a group of people in every society who are disconnected from travel commerce, not curious about what's going on in the world and not likely to possess the disposable income. This segment is a realistic target for the attention of breweries and shops – not cruise line marketers.	7

4 Cruise Trends

The following trends in cruise tourism in the Red Sea were identified:

- According to our survey, the top reason travelers are interested in cruises is the ability to *visit multiple destinations.*

- *More time in port.* There are feature overnight stays, performances in port, and shore excursions, deeper connections with local people and cultures.

- *The cruise industry has not been spared from quality and safety controversies linked to the COVID-19 pandemic*. Given the current impacts that the pandemic has on cruising, it may take some time before the industry recovers and returns to a post-COVID-19 normal. Also, cruise liners will be expected to work more closely with public health authorities worldwide and CLIA (Cruise Lines International Association) to enforce health requirements.

- *Sustainability matters*, and cruise lines are responding to this travel transformation by implementing more environmentally friendly technology and programs to support the destinations they visit. By 2030, the cruise industry has committed to reducing its fleet-wide rate of CO^2 emissions by 40 percent, reports the Cruise Lines International Association.

- *Micro-Cruises.* It's exactly what it sounds like: short, close-to-home jaunts for travelers with little time, but a lot of desire to get away.

- *Solo Travel.* Travelers today are not afraid to venture out by themselves. They want the freedom to travel alone. This trend is only expected to apply to cruises as well as other forms of travel. Women are making more money than ever before. One place where women are putting their money is that of travel. Cruises are a huge part of this travel plan for many women. A safe and protected space where women can relax as they travel will continue to appeal to women all over the globe as they make their personalized travel plans each year.

- *Work and play* are a fact of life for many cruise lines today.

5 Conclusions

The Red Sea is part of the sea roads between Europe, the Persian Gulf and East Asia, and as such has heavy shipping traffic. The marine environment has long been one of the most attractive settings for tourism. Cruise tourism is one of the fastest growing areas within the world's largest industry. 11 cruise lines now represented in the Red Sea. First Class Cruise Lines and Premium Cruise Lines predominate in the Red Sea.

The Royal Caribbean International is a leader. The Oceania Cruises is an outsider.

MSC Cruises - the most active cruise line in the Red Sea. The cruise route programs in the Red Sea are represented by excursions to the regional historic places of interest. The cruise ships make stops for visiting museums, memorial complexes, historical places of interest. The primary reasons that consumers will be motivated to book a cruise in 2021-2022 and beyond, according to trends, include: good value offered by the cruise lines; love of cruising; get away with friends or alone; family reunions, honeymoons/weddings/anniversaries; try out new sustainable ships; shorter cruises; and «work and play» opportunity.

References

1. CLIA (2020). Cruise Lines International Association. https://www.cruising.org
2. Cruise Industry (2020). News MSC announces Red Sea itinerarie. http://www.cruiseindustrynews.com
3. Red Sea MSC Cruises. (2019). http://www.msccruises.com/gl_en/Cruise-Departures/Find-Your-Cruise.aspx?Reg=RED&ddl=y&p=1&

4. Mullally, L. In search of bargains at sea: Deciphering the deals from cruise line. (2018). https://www.montereyherald.com/2010/11/21/in-search-of-bargains-at-sea-deciphering-the-deals-from-cruise-lines/
5. Klein, R. A. The cost of cruising. (2018). http://www.vancouverobserver.com/life/travel/2010/04/13/cost-cruising
6. Cruise Lifestyles (2018).. Cruise Lines – To Suit Your Style. http://cruisediva.com/cruise_style.htm
7. Duman, T., Mattila, A. S. (2005). The role of affective factors on perceived cruise vacation value. Tourism Management, 26(3), 311-323. https://doi.org/10.1016/j.tourman.2003.11.014
8. Hung, K., Petrick J. F. (2011). Why do you cruise? Exploring the motivations for taking cruise holidays, and the construction of a cruising motivation scale. Tourism Management, 32(2), 386-393. https://doi.org/10.1016/j.tourman.2010.03.008
9. Jones, R. V. (2011). Motivations to cruise: An itinerary and cruise experience study. Journal of Hospitality and Tourism Management, 18(1), 30-40. https://doi.org/10.1375/jhtm.18.1.30
10. Marti, B. E. (1991). Cruise ship market segmentation: a 'non-traditional' port case study. Maritime Policy and Management, 18(2), 93-103. https://doi.org/10.1080/03088839100000033
11. Parola, F., Satta, G., Penco, L., Persico, L. (2014). Destination satisfaction and cruiser behaviour: The moderating effect of excursion package. Research in Transportation Business and Management, (13), 53-64. https://doi.org/10.1016/j.rtbm.2014.11.001
12. Zhao, S., Timothy, D. J. (2017). Tourists' consumption and perceptions of red heritage. Annals of Tourism Research, (63), 97-111. https://doi.org/10.1016/j.annals.2017.01.003
13. Cruise Market Watch (2020).Cruise market segmentation; admirals and stowaways. https://cruisemarketwatch.com

Mathematical modeling of the trajectory of a ship as a control object in local planning

Oleksandr Kupraty[1*]

[1]Odessa National Maritime University, Department "Navigation and Marine Safety", Mechnikova str, 34, 65029, Odessa, Ukraine

Abstract. In the article proposes the decomposition of the global planning task in local planning. It is proposed to combine the segment method and the sector method of division of the circle to construct the ship's trajectory in local planning. Corrective coefficients were selected for the correct geometry of the turning trajectory, combining the segment method and the sector method of constructing the ship's trajectory. The article uses formulas of spherical trigonometry; the trajectory of the turn depends on the rudder angle, which in turn depends on the turning ability of the vessel under the given conditions. In determining the value of the angle of the rudder, the control device must take into account the ship's turning ability in the conditions set, the value of the angle of the turn and the constrained water area. The combination of all factors allows to differentiate the ship's turning ability such as: high HG, middle MD or low LW with regard to passage area. The ME shifting operating modes matrix proposed in the article works as a filter of modes of operation and is completely dependent on the readings of pressure, temperature and vibration sensors. The ship's trajectory is constructed using calculations in MS Excel and graphic simulations in the MATLAB environment.

1 Introduction

The development of the modern transport industry in the world is characterized by the intensification of the use of unmanned vehicles. is also the case in the water transport industry. An important feature is the need to increase the ability of such vessels, to expand the range of tasks. The article is devoted to the processing of the task of global planning in local planning when programming a control device, which will control the vessel, according to the requirements for unmanned vessels [1] connection with the above, the topic of the work seems relevant.

One of the important tasks of commercial operation of the vessel is to ensure that the cost of a highly skilled command force is reduced by increasing the level of automation of the ship's control.

The proposed mathematical model will save the cost of crew on merchant ships and reduce the loss of human resources on military vessels during combat operations.

* Corresponding author: kupraty@ukr.net

The decomposition of the global planning mission to simulate the ship's trajectory is an important step towards the creation of an unmanned vessel with a fourth degree of autonomy, according to the classification MARITIME AUTONOMOUS SURFACE SHIPS (MASS) [1]. The fourth degree of autonomy involves making decisions by a management device without human involvement.

2 Analyzing literature and formulating of problem

The choice of literature was based on three principles: relevance, capacity, truthfulness. In order to introduce development into the marine fleet, it was necessary to analyze: the legal acts of the Maritime Safety Committee; global planning task for constructing of ship's trajectory; foundations of discrete mathematics; geometry and cinematic of process; hydrodynamic features of the ship's movement; existing developments on automated (autonomous) ship control, assess their effectiveness; methods for solving direct and reverse geodesic problems; research of engine diagnostics and monitoring.

The regulatory framework for the programming of the control device is set out in the research [1, 3–6]. Modern developments should focus on the report of the Maritime Safety Committee at its 100th session, MSC 100/20/Add.1, Annex 2 [1]. At the source [1] was assessed the degree to which the existing regulatory framework under its purview may be affected in order to address MASS operations.

In source [2] was represented mathematical model of global planning tasks using discrete logic. A problem for ship motion global planning [2] was to establish a link between the top of the graph and the corresponding scenario, avoiding a conflict of priorities when performing a global planning assignment in local planning. Papers [3-6] used as basic international regulatory for Autonomous vessel creation.

Autopilot system known from Japanese patent [7] allows you to adapt the work of the ship's steering gear to the environment by changing control mode. This patent, based on weather data, changes the parameters of the steering machine, but requires constant human control on board. From a Chinese patent application [8] known way to optimize the route of an unmanned vessel based on environmental information. Chinese elaboration is similar to the elaboration [7], but creates the recommended trajectory. This elaboration, like the previous one, has limited functionality and cannot provide the 4th level of autonomy of the vessel [1].

Methods proposed in research [7] and [8] have a significant disadvantage, which does not allow to name the vessels with such devices autonomous: these devices do not classify operational scenarios in real time or form a global planning mission. In research [9] is the development of an unmanned boat called «Surface robot». It is a water robot that does not require a human presence to work and can perform its task autonomously without remote control. It is more intended for dangerous or routine and repetitive military and civilian tasks such as military strikes (respectively with the size of a vessel), geomorphology, patrolling, rescue operations and environmental monitoring, but cannot be used to transport cargoes and people.

Therefore, functionality of existing unmanned vessels is not sufficient to execute difficult tasks for merchant fleet. The above models cannot allow a large unmanned vessel to moor without tugs. Elaboration, mentioned in article, will admit increasing displacement of merchant unmanned vessels and dimensions of military unmanned vessels (their hit power). Research [10-12] dedicated to accidents on fleet. According to statistics mistakes in people's work are the causes 75%–80% of all accidents [10–12]. Human errors are errors that occur when receiving and converting information, when making and executing a decision. The possibilities of discrete mathematics have been researched in paper [13]. In this book were represented theoretical principles of discrete mathematics, which were used in this research.

The theoretical justification for considering the vessel as a control facility is the sources [14,16–19].

In the research [14] authors define a person as a control device, a person's conceptual separated from the means of steering, and the vessel accepted as object of control. Also in work [14] pay high attention to manoeuvring characteristics of the vessel, their calculation and comparison with experiment. «The integral control system includes a control body (control device), controls (steering machine, engines, bow thruster, surface and underwater surveillance equipment, etc.) and a control object (ship)» [2].

In thesis [15] authors noted the main difference between program and device.

In the article [16] were proposed program code for calculating and for constructing the ship's turning circle in MATLAB.

In research [17] were elaborated empirical formulas have been to construct the ship's turning circle and to calculate the geographical coordinates of the ship's center of gravity during turning. In work [18] were proposed method of assessing the reliability of a person's work as a control device on a ship. In work [19] describes the use of sensors to construct a path trajectory. Work [20] is manual "Basics of automation" by Józef Lisowski contains basic information in the field of automation given in a popular-scientific way.

In research [21] describes divergence in constrained conditions, the means of automated radar laying, organizing the work of the bridge team in the control of the movement of the ship. Research [22] contains cinematic models of robotic systems and control algorithms. Paper [23] contains international standards of minimum ITF wages for crew salaries.

Research [24] highlights the main provisions of hydromechanics. In work [25] describes geodesic solutions. Research [26] describes the sector method of constructing a trajectory.

In research [27] illuminated method of calculating distance on rhumb line. In research [28] illuminated method of constructing a trajectory of the turn with the help of two segments. Research [29] it's an internet page dedicated to solving geodesic problems.

The paper [30] presents algorithms for marine diesel engines' working process monitoring designed by analyzing data obtained with IMES pressure sensors (monitoring and diagnostic of main engine).

3 The aim and tasks of elaboration

The aim of the elaboration is to process the global planning task in local planning by the control device, taking into account the manoeuvring characteristics of the vessel, the conditions of navigation and the type of ship power plant.

In order to achieve this aim the following tasks have been identified: to elaborate an algorithm for choosing the rudder angle depending on the ship's turning ability; to select the correcting coefficients of the turning trajectory, combining the segment method and the sector method of constructing the ship's trajectory; to show a graphic simulation of the ship's trajectory in local planning using a programming environment MATLAB; to elaborate mode switching matrix when main engine is braking.

4 Materials and methods used for elaboration

This article shows a mathematical model for procession a global planning task in local planning. When constructing a trajectory, the method of segments was combined [28] with sector method [26] based on the ship's manoeuvring characteristics, such combination is innovation. At the same time, special corrective coefficients were calculated and described in article. When choosing the rudder angle, control device program takes into account external conditions, affecting the steering of the vessel, as well as a safe corridor for manoeuvring. The ship's braking occurs when the appropriate scenario appears, the coordinates of the ship's stop are based on data from beacons, pilot-operator and all surface and underwater surveillance equipment available on the vessel, including radar and optical. The alongside vessel's motion scenario S4 is also calculated and implemented in local planning based on data from beacons, pilot-operator and all surface and underwater surveillance equipment available on the vessel.

The main advantage of this mathematical model is the simplicity of algorithms and the lack of subjective perception. The ship's trajectory is constructed using calculations in MS Excel and graphic simulations in the MATLAB environment.

Here are the definitions for this article:

– local planning – generating the ship's trajectory in the form of a series of points, placed in the route matrix based on the different parameters of the vessel and data from all means of observation; generated route matrix line based on coordinates of route vertex of global planning, using trigonometry formulas,

– segment method – mathematical method used to calculate the turn route of the vessel, which is to calculate the distance between the starting point of the turning and the vertex of the turn, as well as the distance between the top of the turn and the end of the turn, is applied when it allows the distance between the tops of the turns,

– sector method – mathematical method used to calculate the route of the ship's, consisting in dividing the circle of the turn on the arc of a particular sector step and calculating the coordinates of the points separating the circle,

– a set of navigational data is a data with automatic updates over the Internet.

5 Results of elaboration and discussion

5.1 Elaboration of algorithm to select the rudder shift, depending on the ship's turning ability

In the initial form of the path matrix (Fig. 1.) the ship would move from point A to point B without any scenarios, but knowing the purpose of arrival to the point **B** and/or the mooring board, the control device will determine the Fig. 2 scenario, which represents the path matrix with active braking at point **B**.

-	A1	A2	A3	A4	A5	B
A1	0	1	0	0	0	0
A2	-1	0	1	0	0	0
A3	0	-1	0	1	0	0
A4	0	0	-1	0	1	0
A5	0	0	0	-1	0	1
B	0	0	0	0	-1	0

Fig.1. Initial path matrix.

-	A1	A2	A3	A4	A5	B
A1	0	1	0	0	0	0
A2	-1	0	1	0	0	0
A3	0	-1	0	1	0	0
A4	0	0	-1	0	1	0
A5	0	0	0	-1	0	1
B	0	0	0	0	-1	-1

Fig.2. Path matrix with active braking in p. **B.**

But if the braking scenario is calculated earlier, the vessel should stop before the vertex of the corresponding turn no closer than at the point **H**, calculating the coordinates of the stop. According to the segment method [28], segment **MH** will depend on the rudder's shift at the point **H**. In determining the value of the angle of the rudder, the control device must take into account the ship's turning ability in the conditions set (weather conditions, cargo condition of the vessel), the value of the angle of the turn and the constrained water area. The combination of all factors allows to differentiate the ship's turning ability such as: high **HG**, middle **MD** or low **LW** with regard to passage area.

Below is mentioned Table 1, which shows the typical distribution of rudder shifts depending on the turning ability and angle of turn for the vessel with a metacentric height, allowing to put the helm on board. The first column of the table is the assessment of turning ability, the second is the angle of the turn **Q**, the third is the rudder angle **δ**.

Table 1. Typical distribution of rudder shifts depending on the turning ability and angle of turn for the vessel with a metacentric height

Turning ability	Turn angle **Q**	Rudder angle **δ**
HG	$0<Q1\leq 20^{\circ}$	$\delta1=5^{\circ}$
	$20^{\circ}<Q2\leq 30^{\circ}$	$\delta2=10^{\circ}$
	$30^{\circ}<Q3\leq 40^{\circ}$	$\delta3=15^{\circ}$
	$40^{\circ}<Q4\leq 50^{\circ}$	$\delta4=20^{\circ}$
	$50^{\circ}<Q5\leq 55^{\circ}$	$\delta5=25^{\circ}$
	$55^{\circ}<Q6\leq 60^{\circ}$	$\delta6=30^{\circ}$
	$60^{\circ}<Q7$	$\delta7=35^{\circ}$
MD	$0<Q1\leq 10^{\circ}$	$\delta1=5^{\circ}$
	$10^{\circ}<Q2\leq 20^{\circ}$	$\delta2=10^{\circ}$
	$20^{\circ}<Q3\leq 30^{\circ}$	$\delta3=15^{\circ}$
	$30^{\circ}<Q4\leq 40^{\circ}$	$\delta4=20^{\circ}$
	$40^{\circ}<Q5\leq 45^{\circ}$	$\delta5=25^{\circ}$
	$45^{\circ}<Q6\leq 55^{\circ}$	$\delta6=30^{\circ}$
	$Q7>55^{\circ}$	$\delta7=35^{\circ}$
LW	$0<Q1\leq 5^{\circ}$	$\delta1=5^{\circ}$
	$5^{\circ}<Q2\leq 10^{\circ}$	$\delta2=10^{\circ}$
	$10^{\circ}<Q3\leq 20^{\circ}$	$\delta3=15^{\circ}$
	$20^{\circ}<Q4\leq 30^{\circ}$	$\delta4=20^{\circ}$
	$30^{\circ}<Q5\leq 40^{\circ}$	$\delta5=25^{\circ}$
	$40^{\circ}<Q6\leq 45^{\circ}$	$\delta6=30^{\circ}$
	$Q7>45^{\circ}$	$\delta7=35^{\circ}$

An important role in the manoeuvring of the vessel is executed by the safety corridor, which is selected from the database for each section of the track. The section of the track for the safety corridor is not tied to the turning points. The section of the track is calculated percent of the total length of the track, and each section has its own safety corridor. Fig.3. and Fig.4. represent the turn of the ship at a certain turn **Q** for rudder shift.

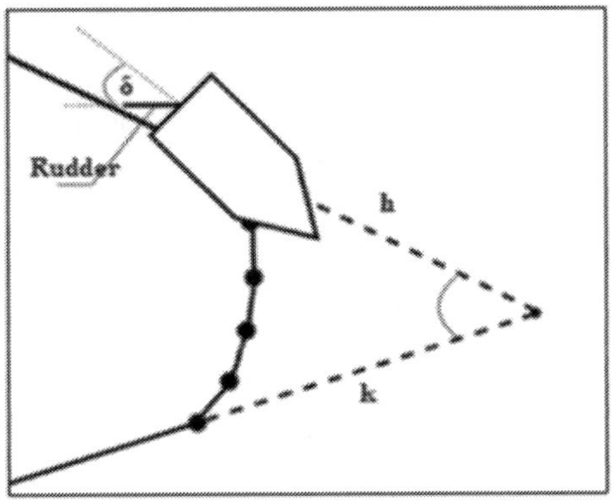

Fig.3. Rudder angle.

Figure 3 shows vessel turn after rudder shift.

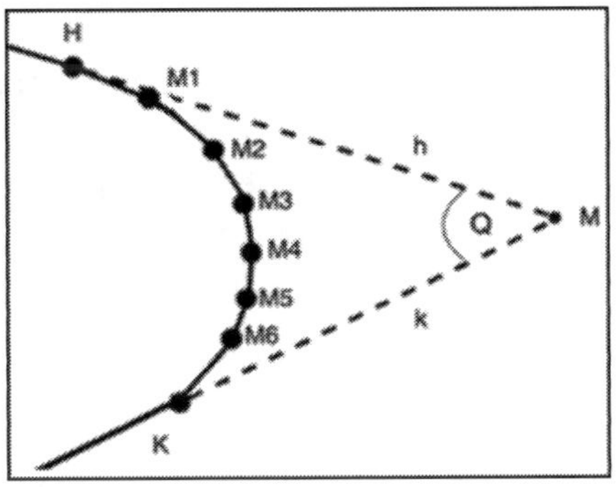

Fig.4. Constructing a turn trajectory.

Figure 4 shows the correction of the first two segments of the turn angle and the final section of the turn angle.

5.2 A selection of correcting coefficients for geometry of the turn trajectory, combining the segment method and the sector method of constructing the ship's trajectory

Length of circular arc **HM1** and **M1M2** corrected by special compensatory coefficients, that compensate for the part of the evolutionary period during which the course does not change or change, but not significantly. The length of the first two arcs indicates $\boldsymbol{d_1}$, $\boldsymbol{d_2}$. The lengths of the other arcs can be defined as $\boldsymbol{d}$. The main coefficient for all arcs is the attrition coefficient $\boldsymbol{k_{di}}$, as the helm is shifted when the course is stabilized in the opposite to the turn direction. This coefficient depends on the initial rudder shift. Coefficients $\boldsymbol{k_{di}}$ and $\boldsymbol{k_{turn1}}$ used for all arcs (Fig. 5, algorithm 1), the rest of the coefficients are only used for $\boldsymbol{d_1}$, $\boldsymbol{d_2}$. Through numerous graphic simulations, the coefficient values $\boldsymbol{k_{di}}$ were calculated for the vessel «UMM QARN» IMO: 9732333, which are presented in Table 2, and other compensating coefficients for the ship «UMM QARN» IMO: 9732333, selection algorithms of which are presented on Figure 5. Coefficients $\boldsymbol{k_h}$ ($\boldsymbol{k_h^{BL}}$) (Fig. 5, algorithm 2)

– these are coefficients dependent on the cargo condition of the ship and the rudder shift **δ**. If the vessel is in ballast, the coefficient $\boldsymbol{k_h^{BL}}$ will be less, and in load condition – $\boldsymbol{k_h}$ will be greater. This is due to the fact that when the ship is in load condition, the turning circle advance increases and the tactical diameter decreases. And when the ship is in ballast condition, turning circle advance decreases, and the tactical diameter is growing. Coefficient $\boldsymbol{k_{turn2}}$ (Fig. 5., algorithm 3) applies at small angle of the turn, as the distance between the top of the turn and its end is less than two length of the vessel. Coefficient $\boldsymbol{k_{dist}}$ (Fig. 5, algorithm 4) is used, if the distance **S** between the tops of the turn is smaller or equal to the differentiated distance for the vessel and the direction of the turns equally (two consecutive courses increase/decrease). Initial formulas for segments **HM** and **MK** [28].

Differential values of turning angles in the questions on the figure 5, (algorithm 1, 3) – these are the angles that the ship's trajectory simulation was oriented towards. «UMM QARN» IMO: 9732333. Algorithm questions (2) on the figure 5 – this is the result of simulations on the main rudder shifts of the ship «UMM QARN» IMO: 9732333.

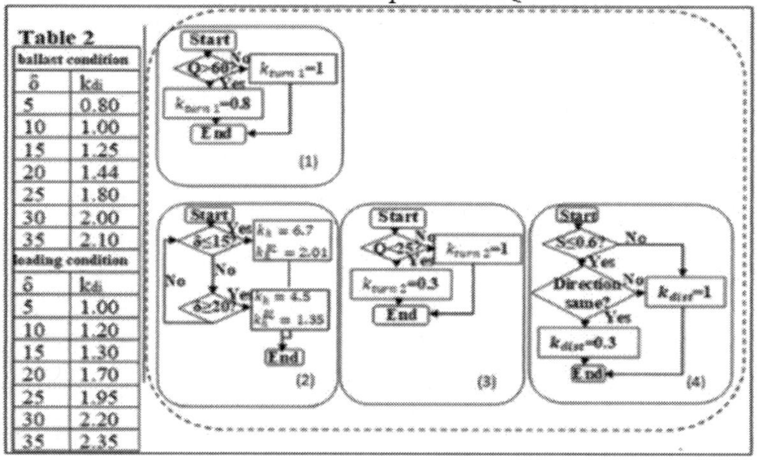

Fig. 5. Algorithm for the selection correcting coefficients.

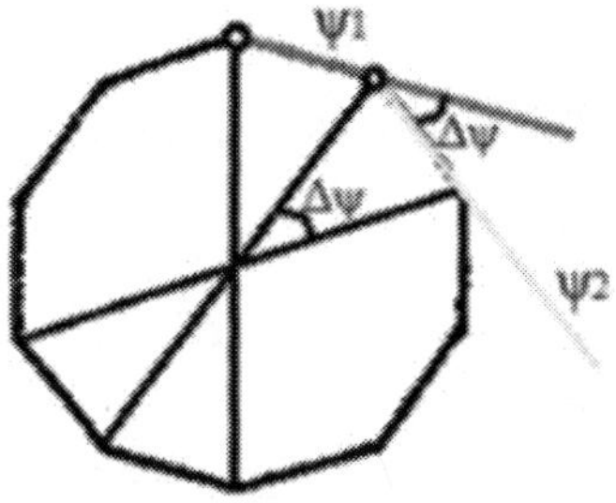

Fig. 6. Change of course during ship turn and sector step.

Length of arc d_0 without the correction would have been calculated according to the formula:

$$d_0 = D_t \times \pi / \Delta\psi' \quad (1)$$

Δψ – this is the value of the exchange course between the two points of the turn arc and at the same time the sector step (Fig.6); Δψ′ – it's an attitude 360°/ Δψ;

These co-components were also successfully tested during a mathematical experiment for the vessel «MSC EMMA» IMO 9463047, that has different turning circle parameters. (Fig. 9.)

Based on all of the above, empirical formulas have been created to calculate the latitude coordinates of points **M1**, **M2**, **M3-Mn** (2,3).

$$Lt_{\{1,2\}} = Lt_{\{H,1\}} + k_h \times k_{dis} \times k_{turn1} \times k_{turn2} \times k_{di} \times d_0 \times \mathrm{COS}(TC_i \times \pi/180)/60 \quad (2)$$

TC_i – true course that is changing by the size of the sector step.

$$Lt_{\{3,4,...n\}} = Lt_{\{2,3,...,n-1\}} + k_{turn1} \times k_{di} \times d_0 \times \mathrm{COS}(TCi \times \pi/180)/60 \quad (3)$$

And longitude should be considered by the formula:

$$Ln_{n+1} = Ln_n - (7915.70447 \times \lg[tg((45+Lt_n/2)\times\pi/180)\times((1-e\times \mathrm{SIN}(Lt_n\times\pi/180))/(1+ +e\times \mathrm{SIN}(Lt_n\times\pi/180)))]-7915.70447\times\lg[tg((45+Lt_{n+1}/2)\times\pi/180)\times((1-e\times \mathrm{SIN}(Lt_{n+1}\times\pi/180))/ /(1+e\times \mathrm{SIN}(Lt_{n+1}\times\pi/180)))])\times tg((TCi\times\pi/180))/60 \quad (4)$$

Thus, the trajectory is generated as a chain of points, the coordinates of which are calculated by trigonometry. Segment method [28] snaps the calculation of the segment value **MK** to turning circle transfer $\boldsymbol{L_2}$, that necessitates its adjustment by changing the formula when programming trajectory parameters. Calculation of **MH** tied to the advance of turning circle $\boldsymbol{L_1}$, that's why at small angle of the rudder shift and small angles of the turn it should also be corrected.

Based on the above-mentioned, it can be argued that the circumference of each turn has its own geometric features, but these features are not only in the evolutionary period of the turn, but also in its completion. The trajectory of the ship turn can be called an ellipsoid, as its radius changes. In local planning of the ship's turns, it is necessary to solve the problem of entering the turn trajectory to the point **K.** This task is solved by successive experimental approximations. When solving this task, it is important to determine when the turn trajectory should be simplified into a straight line to the point **K.**

For vessel «UMM QARN» IMO: 9732333 it was definitely that if the angle of the turn **Q** more than 45°, the trajectory is simplified into a straight line and enters the point **K**, when remains 0.33 of the difference between the initial and the finish course until the turn is completed. Under other conditions, the trajectory is simplified into a straight line and enters the point **K**, when remains 0.36 of the difference between the initial and the finish course (on figure **4** this transition is shown by a segment **M6K).**

As you know, the main task of local planning is to change the trajectory movement of the ship in real time. Consider the implementation of passive and active braking in local planning, where scenario **S1** – it's a passive braking scenario, a **S2** – it's an active braking scenario [2] (Fig.7 a, b). At the same time, the need to accelerate or slow down the ship is not a scenario, as it does not change the route matrix. The passive braking scenario is determined numerically if the speed required for safe movement is below the minimum speed at which the vessel is steered. the scenario of active braking is determined if a full stop of the ship is needed. If you should stop before turning at a minimum distance from the top, the calculation of geographic coordinates uses bearing and distance to the point **H**. Formula 5 and 6 are used to calculate coordinates for passive braking at the point H in another more distant from the top turning point H', H'', H''' and etc.

$$LT_H = LT_M + \mathrm{dist.}\times\cos((PL_H)\times\pi/180)/60 \quad (5)$$

$$LN_H = LN_M - (7915.70447\times\lg(tg((45+LT_M/2)\times\pi/180)\times((1-e\times\sin((LT_M)\times\pi/180)/(1+ +e\times\sin((LT_M)\times\pi/180))))) - 7915.70447\times\lg(tg((45+LT_H/2)\times\pi/180)\times((1-e\times\sin((LT_H)\times\pi/180)/ /(1+e\times\sin((LT_H)\times\pi/180))))))\times tg((PL_H)\times\pi/180)/60 \quad (6)$$

When the active braking scenario is running, which simultaneously allows you to move astern (and the distance you need to walk back is calculated in meters), formula for calculating latitude will be as:

$$LT_H = LT_M + (\text{dist.} + dist._m/1852)\times\cos((PL_H)\times\pi/180)/60 \quad (7)$$

Scenario S3 – this is the scenario of skipping a planned point (points).

5.3 Graphic simulation of the ship's trajectory in local planning using a programming environment MATLAB

On figure 7(b) represented active braking scenario in point 13. When an point falls out of the route matrix in local planning, the trajectory is based on the vertex that remain (Fig. 8 a, b).

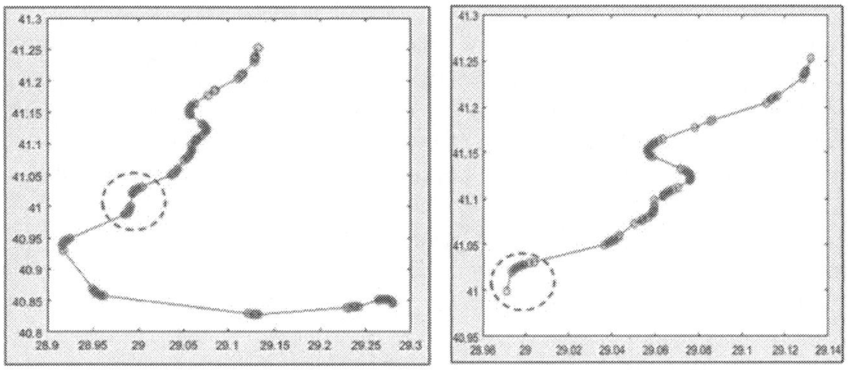

a *b*

Fig. 7. Constructing a trajectory without braking and braking at the vertex 13.

Slowing or accelerating is not part of the scenario matrix because the velocity change does not change the route matrix. This article only looks at the deviation from the course through a point pass, scenario 3. The usual deviation from the course in this article is not considered, as this is a separate scenario of local planning not related to global planning. In developing such a deviation from the course should be calculated not only the possibility of deviation from the course, but also the possibility of returning to the course.

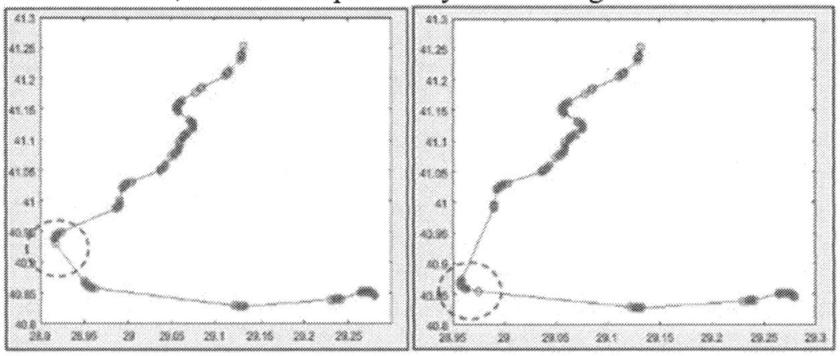

Fig. 8. Trajectory of moving with skipping and without skipping vertex 14.

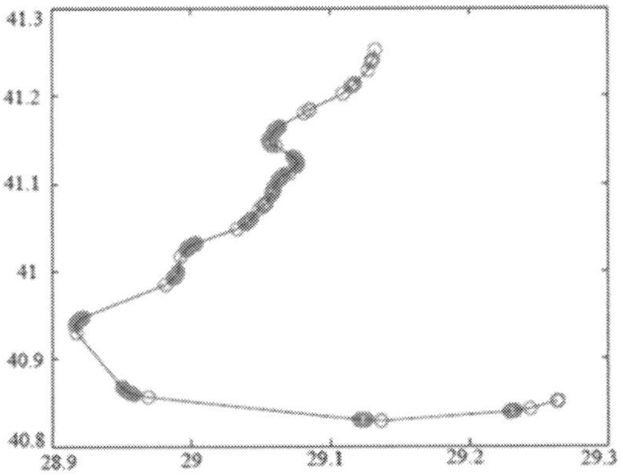

Fig.9. Constructing the ship's trajectory MSC EMMA using corrective coefficients.

The trajectory of the turn can be presented as a matrix, Fig. 10.

$$\left[\begin{vmatrix}\varphi_H\\ \lambda_H\end{vmatrix},\begin{vmatrix}\varphi_1\\ \lambda_1\end{vmatrix},\begin{vmatrix}\varphi_2\\ \lambda_2\end{vmatrix}\dots\dots\dots\begin{vmatrix}\varphi_9\\ \lambda_9\end{vmatrix},\begin{vmatrix}\varphi_{\dots}\\ \lambda_{\dots}\end{vmatrix},\begin{vmatrix}\varphi_{\dots}\\ \lambda_{\dots}\end{vmatrix},\begin{vmatrix}\varphi_{\dots}\\ \lambda_{\dots}\end{vmatrix},\begin{vmatrix}\varphi_{i-2}\\ \lambda_{i-2}\end{vmatrix},\begin{vmatrix}\varphi_{i-1}\\ \lambda_{i-1}\end{vmatrix},\begin{vmatrix}\varphi_i\\ \lambda_i\end{vmatrix},\begin{vmatrix}\varphi_K\\ \lambda_K\end{vmatrix}\right]$$

Fig. 10. The ship's turn route matrix.

Scenario S4 – this is the scenario of active braking and alongside movement at the same time. The alogside movement scenario occurs as a result of the pilot's command, or as a result of linear calculations of the control device. Starboard side alongside movement code is "1", port side is "0". The trajectory of the alongside movement is calculated trigonometrically using the exchange angle 90° or -90°, as presented in the formula (8,9).

$$LT_H = LT_M + \text{dist.} \times \cos((PL_H + (\pm 90)) \times \pi/180)/60 \quad (8)$$

$$LN_H = LN_M - (7915.70447 \times \lg(\text{tg}((45 + LT_M/2) \times \pi/180) \times ((1 - e \times \sin((LT_M) \times \pi/180)/(1 + e \times \sin((LT_M) \times \pi/180))))) - 7915.70447 \times \lg(\text{tg}((45 + LT_H/2) \times \pi/180) \times ((1 - e \times \sin((LT_H) \times \pi/180)/(1 + e \times \sin((LT_H) \times \pi/180)))))) \times \text{tg}((PL_H + (\pm 90)) \times \pi/180)/60 \quad (9)$$

In formulas use back bearing PL_H, so when the alongside movement directed to port side, the formula uses (+90), and when to starboard side – use (-90).

Using trigonometry formulas, the control device can also calculate the geographic coordinates of the center of gravity. After passing through the ship axis $\boldsymbol{x}$, can be found out the coordinates of the antenna GPS – receiver $\boldsymbol{X_A}$ and calculate the difference of coordinates in meters on the axis $\boldsymbol{x}$ between the position of the antenna and the position of the ship's center of gravity ΔX_{A_G}. The formulas for calculating geographical coordinates will look like this:

$$LT_G = LT_A + (\Delta X_{A\_G}/1852) \times \cos(PL_{A\_G} \times \pi/180)/60 \quad (10)$$

$$LN_G = LN_A - (7915.70447 \times \lg(\text{tg}((45 + LT_A/2) \times \pi/180) \times ((1 - e \times \sin((LT_A) \times \pi/180)/(1 + e \times \sin((LT_A) \times \pi/180))))) - 7915.70447 \times \lg(\text{tg}((45 + LT_G/2) \times \pi/180) \times ((1 - e \times \sin((LT_G) \times \pi/180)/(1 + e \times \sin((LT_G) \times \pi/180)))))) \times \text{tg}((PL_{A\_G} + (\pm 90)) \times \pi/180)/60 \quad (11)$$

Thus, the control device, using GPS receivers on a large vessel, will be able not only to receive information about the coordinates of the stern, bridge or bow due to the position of GPS antennas, but also will be able to calculate the coordinates of the center of gravity of the ship.

5.4 Elaboration a main engine shifting operating modes matrix

Scenario S1 is the scenario of passive bracking, S2 is the scenario of passive bracking, S4 is the scenario of active braking and side movement. Braking on vessels with the fourth level of autonomy should be carried out without an operator and without the constant presence of service personnel in the engine rooms.

In passive and active braking, the control device must monitor the main and supportive mechanisms of the engine room. If necessary, the control device should turn off heating devices of automated boilers, fuel pumps, engine room fans. Emergency warning alarm system must be directly connected to the control device.

In order to monitor the safety of braking, the vessel is proposed to use the following main engine shifting operating modes matrix (Fig. 11).

	Temperature of cylinders	Pressure of gases inside cylinders	Vibration on ME flywheel
FA - FULL AHEAD (Manoeuvring mode)	0	0	0
HA – HALF AHEAD	0	0	0
SA – SLOW AHEAD	0	0	0
DSA – DEAD SLOW AHEAD	0	0	0
STOP ENGINE	0	0	0
DAS – DEAD SLOW ASTERN	0	0	0
SAS – SLOW ASTERN	0	0	0
HAS – HALF ASTERN	1	0	1
FAS (mm) – FULL ASTERN Manoeuvring mode			

Fig. 11. Main engine shifting operating modes matrix.

In this matrix, transitions between modes occur as follows: if the row is completely zero, an indication from all ME sensors is normal–the active braking continues, if is not – the control device expects all indicators to return to normal before switching the next main engine mode or returning to previous position or to the position STOP ENGINE. On Fig. 11. mentioned situation, when need to return to previous ME operating mode in case of temperature and vibration.

The temperature of the cylinder is determined by three indicators: temperature, cooling water temperature, oil temperature. Gas temperature sensors are on exhaust manifold pipes. Cooling water temperature sensors are located near the engine cylinder bushings.

Crankshaft bearing temperature sensors indirectly indicate the temperature inside the cylinder. Indications of the oil temperature sensor at the exit of the engine are also important in terms of diagnosing the entire engine.

The pressure of the gases inside the cylinders is determined by pressure sensors on the cylinder covers [30]. The level of vibration is determined by stationary sensors on the engine flywheel.

6 Conclusions

1. To perform global planning scenarios, an algorithm has been developed to select the rudder angle. The algorithm gives steering option to respond to real-time environmental changes. The main advantage of this mathematical model is the simplicity of algorithms and the lack of subjective perception.

2. By connecting the segment method and the sector method and correcting coefficients

the trajectory of turns is constructing correctly.

3. Main engine shifting operating modes matrix is an effective tool for monitoring and manage the work of ME.

4. The practical usefulness of this elaboration is shown through a graphic simulation. From source [2] it is known that the relative reliability of the device based on this development is 1.33, and economic efficiency 37626$ for crew 23 persons per month on one merchant ship according to ITF ILO Minimum Wage Scale [23]. On warship, this elaboration can save many lives while performing combat missions in a priori unequal conditions. The proposed mathematical modeling of the ship's trajectory in local planning brings the creation of a control electronic device that will be able to control large vessels. At the same time, the control device will operate in accordance with international requirements set out in the sources [1, 3-6].

References

1. IMO. Framework for the Regulatory Scoping Exercise for the Use of Maritime Autonomous Surface Ships (MASS). MSC 100/20/Add.1, Annex 2. 2018.
2. Kupraty, O.G. (2020). Mathematical modeling trajectory of a ship as a control object in global planning. Transport development, 1(6). 40-55.https://doi.org/10.33082/td.2020.1-6.04
3. IMO. Code on Alerts and Indicators. (2009). A.1021(26).
4. IMO. Convention on the International Regulations for Preventing Collisions at Sea, 1972 (COLREGSs), consolidated edition 2018, amended in 2010, 2016.
5. IMO. Revised Maritime Policy and Requirements for a Future Global Navigation Satellite System (GNSS).(2001). A.915(22).
6. IMO (2014). SOLAS Consolidated Edition, 2014 (IF110E).
7. Yokoyama, Akira. (1981). Control system for unmanned sailing ship. https://worldwide.espacenet.com
8. Ding Fuguang, Liu Xiangbo, Wang Chenglong, Wang Yuanhui, & Zhang Zhaoqing (2018). Unmanned ship route optimization method based on environment forecast information.
9. Chen Qichang, Huang Zejia, Lin Yuanping, Liu Peiheng, & Ouyang Quanjie (2015). Cruise control method and system of unmanned ship. https://worldwide.espacenet.com/patent/search/family/054714755/publication/CN105129064A?q=CN105129064
10. Katsman, F.M. (2006). Avarijnost morskogo flota i problemy` bezopasnosti sudokhodstva. Transport Rossijskoj Federaczii, 5, 82-84
11. Liberman, A.N. (2006). Tekhnogennaya bezopasnost: chelovecheskij faktor. SPb.: Izdvo VIS.
12. Rodriguez, J.L. (2011). Legal gaps relating to labour safety and health in the maritime transport sector in Spain. Int. Marit.Health, 62(2), 91-97.
13. Galiev, Sh.I. (2005). Diskretnaya matematika. Kazanskij Gosudarstvenny`j tekhnicheskij universitet imeni I. N.Topoleva, Kazan.
14. Demin, S.I., Zhukov, E.I., & Kubachev, N.A. (1991). Upravlenie sudnom. Izd. Transport. Moskwa.
15. Petrov, I. M., Rudnichenko, N. D., Kupraty, O. G. (2020). Specificity Of Using Modern Information Technologies In Service Ergatic Systems On Marine Transport. XXXIX International Scientific and Practical Internet Conference Modern vector of

science development, 19-22.

16. Kupraty, O. (2020). Implementation of the algorithm for calculation course (bearing) on rhumb line and constructing the trajectory of the ship's turning circle in the MATLAB programming environment. The scientific heritage. (Budapest, Hungary), 1,(60), 40—45. https://doi.org/10.24412/9215-0365-2021-60-1-40-45
17. Kupraty, O.G. (2020). Elaboration of empirical formulas for constructing the trajectory of the vessel's turning circle and calculating the geographical coordinates of the vessel's gravity center when turning. The 10th International scientific and practical conference "Priority Areas of Science Research. Primedia E-launch LLC, USA, Washington. 14-21.
18. Kupraty, O.G. (2020). Nadi`jni`st` roboty sudnovodi`ya yak upravlyayuchogo pristroyu na sudni. The 3rd International scientific and practical conference, - World science: problems, prospects and innovations, 669-771.
19. Garbacz, M. (2006). Planowanie ścieżki dla robota mobilnego na podstawie czujników odległościowych" Automatyka 2006, 10(3), Akademia Górniczo-Hutnicza, Krakow.
20. Lisowski, J. (2015). Podstawy automatyki, Akademia Morska w Gdyni.
21. Maltsev, A.S. (2005). Manevrirovanie sudov pri raskhozhdenii, Odessa.
22. Tchoń, K., Mazur, A., Hossa, R., Dulęba, I., Muszyński, R. (2000). Manipulatory i roboty mobilne, Modele planowanie ruchu sterowanie. Robotyka, Warszawa..
23. International Transport Workers' Federation (2020). ITF ILO Minimum Wage Scale Using Joint ITF/ISF Interpretation of the ILO Recommended Minimum Wage for an AB - extrapolated on basis of ITF Standard Agreement Differentials Rates applicable from 1st January 2020.
24. Chizhiumov, S.D. (2007). Osnovy` gidrodinamiki: uchebnoe posobie GOUVPO «KNAGTU».
25. Morgas, W., & Kopacz, Z. (2013). Rhumb-Line Sailing by compution. VERSITA, Reports on Geodesy, 94, 14 –26. https://doi.org/10.2478/rgg-2013-0003
26. Ślączka, W., & Guziewicz, J.(1997). Metody wyznaczania obszaru manewrowania statku stosowane w badaniach symulacyjnych. VII Międzynarodowa Konferencja Naukowo-Techniczna Inżynierii Ruchu Morskiego. Szczecin. 125-142.
27. Tijardović, I. (2000). Rhumbline Distances. The Journal of Navigation, 53(1), 187-191.
28. Vilskiy, G. B, Maltsev, A.S., Bezdolny, V.V., Gonczarov, E.I. (2007). Navigaczionnaya bezopasnost` pri loczmanskoj provodke sudov. Odessa-Nikolaev: Feniks.
29. GIS-Lab (n.d.). https://gis-lab.info/qa/angles-rhumb.html
30. Neumann, S, Varbanets, R, Kyrylash, O, Maulevych, V, & Yeryganov, O. (2019). Marine diesels working cycle monitoring on the base of IMES GmbH pressure sensors data. Diagnostyka,20(2). 21-26. https://doi.org/10.29354/diag/104516

Peculiarities of financing investment infrastructure projects in Ukraine

Nataliia Savina[1], *Nataliia* Kovshun[1*], *Alla* Zhemba[1], *Alisa* Ventsuryk[1], and *Oksana* Kliukha[1]

[1]National University of Water and Environmental Engineering, 33028, Rivne, Ukraine

Abstract. The purpose of the article is to show the role of international financial organizations in financing the development of the transportation network of Ukraine. It is shown the necessity of implementation of infrastructure projects for the development of the state economy and increase of its competitiveness. The structure of sources of financing in the development of highways is shown and the gradual reduction of the share of investments at the expense of the State budget and the insignificant role of public and private partnership in the transport sphere are revealed. The cooperation of Ukraine with international financial organizations: the International Bank of Reconstruction and Development, the European Bank of Reconstruction and Development, and the European Investment Bank in the implementation of infrastructure projects have been studied. A retrospective analysis of the available loan agreements was performed. The expected positive consequences of international investment for the transport sector of Ukraine and the economy as a whole have been identified. Information is given on real investment infrastructure projects with the participation of international financial institutions and the financial conditions of their implementation. The shifts in the arrangement of highways of Ukraine at the expense of international investments are listed.

1 Introduction

To overcome the difficult economic situation, Ukraine needs to modernize the road network, which can be done through the implementation of investment infrastructure projects. Given the unsatisfactory financial condition of the transport sector in the short term, attracting investment funds will help stabilize the road sector by providing the necessary funds for road maintenance. In the medium and long term – economic growth and competitiveness, by strengthening ties between regions and improving transit corridors [1]. Modernization of the road network will enhance economic activity by reducing travel times and reducing transportation costs between major centers in Ukraine and their regional trading partners.

At the same time, improved road infrastructure will increase productive investment in the territorial communities that benefit from investment infrastructure projects and create

* Corresponding author: nata-an@ukr.net

jobs for their residents, which in turn will increase household incomes and consumption levels, as well as reduce poverty. It should also be borne in mind that the development of transport requires not only the construction of roads but also involves solving problems such as reducing road congestion, environmental protection, traffic safety, etc. [1].

2 Analysis of recent approaches and publications

The resumption of Ukraine's economy and growth of its competitiveness largely depends on the efficiency of the transport system. This relationship has been studied by scientists from different countries and using different approaches [2]. India's experience has shown that there are long-term interrelations between transport infrastructure and economic development, where economic development being the main factor in most cases [3]. The paper [4]describes the concept of transport infrastructure in Latvia as an important part of the state transport system. The impact of infrastructure development on the economic progress of India and China also confirms the importance of building a transport system, including the use of foreign investment [5] . In general, it is introduced the impact of road development on economic growth in the new EU member states [6].

3 Results

In order to restore and develop public roads of national importance for their integration into the European transport system and increase the level of traffic safety, speed, comfort, and efficiency of transportation, the Cabinet of Ministers of Ukraine in 2013 developed and approved the State Targeted Economic Program for Public Roads use for 2013-2018.

It is state support and its quality that determine the development of the transport industry [7]. This program was not implemented in full due to underfunding in 2015-2016. Targeted funding sources were canceled, and the planned expenditures for road development in the general fund of the state budget were usually underfunded by the Treasury. As a result, in 2015-2016, the construction of highways was actually stopped, the volume of work on the current average repair was ten times lower than the norm, the pothole was eliminated late and not on all roads. In 2016-2017, due to increased funding, repair and construction work on public roads were significantly intensified, but the lists of facilities were formed not on the basis of this Program, but by separate orders of the Cabinet of Ministers of Ukraine. This unstable state of transport financing was caused by political instability in Ukraine. It should be noted that political and institutional interference in economic processes must be balanced and justified. Attracting foreign, regional, or private investments in the implementation of infrastructure projects depends on government strategies [8, 9].

The projected amount of funding for the State Targeted Economic Program for the Development of Public Roads of State Importance for 2018-2022 is UAH 372.3 billion, of which 67.5% should be financed from the State Budget.

During the implementation of the program, the amount of funding increases significantly over the years from 30.0 billion in 2018 to 97.5% in 2022 (Table 1).

It should be noted that significant changes in funding sources during the implementation of this program. If in 2018, the amount of funding from the State budget in 2018 is 87.6%, and for other sources, including investors' funds account for 12.4%, then in 2022 - 48.9% and 47.3%, respectively. At the same time, it is not typical for Ukraine to attract private investors, although public-private partnerships for individual investment projects are the most rational [10].

Table 1. Financing of the State Target Economic Program of highways development of public use of the state value for 2018-2022.

Sources of funding	Volume of financing, million hryvnias	Including by years, million hryvnias				
		2018	2019	2020	2021	2022
State budget, including:	251 300	26 328	35 984	95 203	46 132	47 653
Special fund of the state budget	212 424	20 025	35 217	63 395	46 133	47 653
General fund of the state budget	38 877	6 302	766	31 808		
Funds of international financial organizations	20 256	3 722	4 463	4 440	3 910	3 719
Other sources, including investors' funds	100 781		514	18 043	36 051	46 172
Total	372 337	30 050	40 962	117 687	86 094	97 545

Financing schemes for infrastructure projects implemented with the participation of international financial institutions are receiving increasing attention around the world. At the same time, there is a defined range of issues that the government must address in order to obtain the necessary investment [11]. For many years, Ukraine has been actively and effectively cooperating with international financial organizations: the International Bank for Reconstruction and Development (IBRD), the European Bank for Reconstruction and Development (EBRD), and the European Investment Bank (EIB). Their activities, among other things, aim to assist in the implementation of investment infrastructure projects (overhaul, reconstruction, and construction of roads, institutional development), which is carried out through long-term lending and consulting services the implementation of projects.

Implementation of investment infrastructure projects, with the participation of the IBRD, EBRD, and EIB, will cause the following economic effect:

- reducing the number of accidents related to unsatisfactory road conditions;
- reduction of the negative impact of the road transport complex on the environment;
- strengthening of roadsides will allow to eliminate dust formation, will improve drainage from the carriageway and from the road, will reduce erosion of soils and pollution of open reservoirs and ground waters;
- reduction of annual operating costs of road transport, taking into account the loss of passengers as a result of travel,
- reduction of annual investment in road transport;
- reduction of annual unproductive loss of working time of passengers on the road;
- reduction of consumption of motor fuels,

The reduction of annual costs for current repairs of highways resulted from the improvement of transport and operational condition, which can be directed to current repairs and maintenance of other highways [12-14]. According to the cooperation with international financial organizations, in 2018 works and services were financed in the amount of UAH 3.7 billion, in 2019 - UAH 4.3 billion. Institutional development includes the provision of consulting services, training, and supply of equipment to strengthen the traffic management system and improve the maintenance of roads following an international practice.

The purpose of the implementation projects of investment infrastructure projects with the participation of the IBRD is the rehabilitation of the international highway M-03 (Kyiv - Kharkiv - Dovzhanskyi checkpoint), improving road safety (elimination of places of concentration of accidents) and institutional development. With the participation of the EBRD, the M-06 (Kyiv-Chop) highway is being restored and the financing of the road sector is being reformed. With the joint participation of the EBRD and the EIB, the reconstruction of the last section of the M06 road, the improvement of Ukraine's road infrastructure, and the promotion of further reform in Ukraine's road sector (Table 2).

Table 2. International highways of Ukraine, on which investment infrastructure projects are implemented.

Index & № roads	Names of highways	Length, km.	Comments
M-03	Kyiv - Kharkiv - Dovzhanskyi (in the city of Rostov-on-Don); entrances: to Kharkiv International Airport, to Kharkiv, to Poltava	937,3	Is essentially important for Ukraine's integration with the EU road network as a potential transit corridor of national and international importance
M-06	Kyiv - Chop (in the city of Budapest via Lviv, Mukachevo and Uzhhorod); entrances: to Zhytomyr, to Novohrad-Volynskyi, to Lviv, to Uzhhorod, western entrance to Rivne, southern approach to Rivne	918,1	Road of international importance on the territory of Ukraine, which passes through the territory of Kyiv, Zhytomyr, Rivne, Lviv and Zakarpattia regions and is part of the European transport corridors №3 and №5

The first loan agreement was concluded in December 2000 with the EBRD. Currently, 9 credit agreements have been concluded between Ukraine and international financial organizations, including 4 credit agreements with the EBRD (total value of EUR 825.0 million), 2 credit agreements with the EIB (total value of EUR 750.0 million), and 3 loan agreements with the IBRD (total value of $ 1,187.0 million). Information on the implementation of investment infrastructure projects with the participation of international financial institutions and the relevant financial conditions is provided below (Table 3, Table 4, Table 5, Table 6).

Table 3. Information on the implementation of investment infrastructure projects with the participation of the IBRD.

Name	Value
Allocated funds, million USD	1 187,8
Disbursed funds, million USD	853,3
Funds mastered,%	71,8
Concluded contracts	67
including: construction work	25
consulting services	42
Deadline for projects	2009 - 2021
Number and status of projects	3
Put into operation (repaired/built) roads (km.) And man-made structures (pieces)	259,1 km. / 53 pc.
Loan maturity	2009 - 2038

The beneficiary of the agreements is the Ministry of Finance of Ukraine or the State Agency of Motor Roads of Ukraine. The source of repayment is the state budget.

Table 4. Financial conditions for the implementation of investment infrastructure projects with the participation of the IBRD.

Name	Projects		
	RSIP I	RSIP II	RSDP
One-time sales commission	0,25%	0,25%	0,25%
Sub-loan agreement commission	0,50%		
Reservation fee			0,25%
Interest rate	LIBOR + variable margin	LIBOR + variable margin	LIBOR + variable margin
The repayment schedule of the principal amount of the loan	May 15 and November 15, starting from 11/15/14	April 15 and October 15, starting from 15.10.18	February 15 and August 15, starting from 15.02.2022
The term of the loan	2009 - 2038	2012 - 2029	2016 - 2032

Table 5. Information on the implementation of investment infrastructure projects with the participation of the EBRD and the EIB.

Name	Value
Allocated funds, EUR million	1 475,0
including European Bank for Reconstruction and Development	825,0
European Investment Bank	650,0
Funds disbursed, EUR million	1 167,8
including European Bank for Reconstruction and Development	754,9
European Investment Bank	412,9
Funds mastered,%	79,2
including European Bank for Reconstruction and Development	91,5
European Investment Bank	63,5
Number of concluded contracts	37
including construction work	22
consulting services	15
Deadline for projects	2000 - 2021
Number and status of projects	4 Completed - 3, in process - 1
Put into operation (repaired/built) roads (km.) And man-made structures (bridges, overpasses, pedestrians…)	987.7 km. / 169 artificial structures
Loan maturity	2000 - 2031

The main achievements in the implementation of investment infrastructure projects with the participation of IBRD:

- overhaul and reconstruction of 259.1 km were performed. roads, of which 82.7 km of roads were brought from the II to the I technical category;

- the strengthening of the existing pavement by the technology of cold recycling was carried out. For the device of the top layer of a covering the crushed stone-mastic asphalt concrete made with the use of modern additives that will allow overcoming problems of a track in the summer period is used. Additional strengthening of asphalt concrete layers on lanes of heavy transport is provided by a special reinforcing grid;

- to improve winter maintenance and increase road safety, 6 road meteorological stations were installed, 53 man-made structures were built, of which: 32 man-made pedestrian crossings at different levels and 21 bridges, overpasses, and cattle runs; - bringing the M-03 to the highway standard, which was a direct contribution to a significant reduction in the number of accidents in the area between Boryspil and Lubny from 248 per year before the project to 10 per year of project closure.

Table 6. Financial conditions for the implementation of investment infrastructure projects with the participation of the EBRD and the EIB.

Name	Projects			
	№ 3		№ 4	
	EBRD	EIB	EBRD	EIB
One-time commission	1,00%	500,0 тис. євро	1,00%	0,25%
Commitment fee	0,50%	0,10%	0,50%	0,10%
Subcredit Commission	0,50%	0,50%	0,50%	0,50%
Interest rate	EURIBOR+1%	EURIBOR + 0,55%	EURIBOR+1%	EURIBOR+0,75%
The term of the loan	2006 - 2022	2007 - 2031	2010 - 2021	2011 - 2020

It is planned to achieve:

- Rehabilitation and improvement of traffic safety in selected sections of M-03,

- building institutional capacity to facilitate the implementation of reforms in the field of maintenance.

- implementation of road safety and network management system.

Main achievements in the implementation of investment infrastructure projects with the participation of the EBRD and the EIB:

- 2006 - 224 km of roads were put into operation. 52 bridges and overpasses were repaired and/or renovated, {18} - 181 km were commissioned in 2009. 26 bridges and 25 overpasses were repaired and/or renovated,

- 2013 - improvement of road conditions on 341.9 km of repaired road M-06 Kyiv-Chop;

- overhaul of roads on general roads 240.8 km long. 66 artificial structures were repaired;

It is planned to achieve:

- overhaul of roads with a total length of 142 km with bringing 22 km to the I technical category.

- repair of artificial structures (17) and construction of new artificial structures (45).

Implementation of investment infrastructure projects should be subordinated to the task of structural restructuring of the transport sector, aimed at ensuring balanced and harmonious development of all parts of the complex, to renew the production capacity, increase the efficiency of its use. For the further development of the transport industry, it is necessary to concentrate investments on the introduction of high-performance and the latest innovations and information technologies, renewal, and modernization of fixed assets and infrastructure.

4 Conclusions

Ensuring a favorable investment climate remains a matter of strategic importance, the implementation of which depends on the effectiveness of economic growth measures in Ukraine, the effectiveness of investment projects, the possibility of modernization of the transport sector. The formation of a favorable investment climate should take place both at the macro and regional levels, optimizing the whole set of political, legal, economic, social, environmental, organizational, informational, industrial and regional specific factors.

References

1. Freeman, P. N. (2009). Ten years of World Bank action in transport: evaluation. Journal of infrastructure systems, 15(4), 297-304.
2. Candemir, Y., & Çelebi, D. (2017). An inquiry into the analysis of the Transport & Logistics Sectors' Role in Economic Development. Transportation research procedia, 25, 4692-4707.
3. Maparu, T. S., & Mazumder, T. N. (2017). Transport infrastructure, economic development and urbanization in India (1990–2011): Is there any causal relationship?. Transportation research part A: policy and practice, 100, 319-336.
4. Skorobogatova, O., & Kuzmina-Merlino, I. (2017). Transport infrastructure development performance. Procedia Engineering, 178, 319-329.
5. Kim, M. J., & Nangia, R. (2010). Infrastructure development in India and China: A comparative analysis. In Physical infrastructure development: Balancing the growth, equity, and environmental imperatives (pp. 97-139). Palgrave Macmillan, New York.
6. Rokicki, B., & Stępniak, M. (2018). Major transport infrastructure investment and regional economic development–An accessibility-based approach. Journal of Transport Geography, 72, 36-49.
7. Kyriacou, A. P., Muinelo-Gallo, L., & Roca-Sagalés, O. (2019). The efficiency of transport infrastructure investment and the role of government quality: An empirical analysis. Transport Policy, 74, 93-102.
8. Cerra, V., Cuevas, A., Goes, C., Karpowicz, I., Matheson, T., Samake, I., & Vtyurina, S. (2017). Determinants of Infrastructure and Its Financing. Emerging Economy Studies, 3(2), 113-126.
9. Tei, A., & Ferrari, C. (2018). PPIs and transport infrastructure: Evidence from Latin America and the Caribbean. Journal of Transport Geography, 71, 204-212.
10. Makarov, I., & Plotnikov, V. (2018). Public-private partnership as a financial instrument for the transport infrastructure modernizing. In MATEC Web of Conferences (Vol. 170, p. 05012). EDP Sciences.
11. Kumaraswamy, M., & Zhang, X. (2001). Governmental role in BOT-led infrastructure development. International Journal of Project Management, 19(4), 195-205.
12. Kvach, Y., Piatka, N., & Koval, V. (2020). Management of sustainable entrepreneurship adaptation to tax changes in environmental investment. Baltic Journal of Economic Studies, 6(5), 96-105. https://doi.org/10.30525/2256-0742/2020-6-5-96-105
13. Arsawan, I. W. E., Koval, V., Duginets, G., Kalinin, O., & Korostova, I. (2021). The impact of green innovation on environmental performance of SMEs in an emerging economy. In E3S Web of Conferences, 255, 01012.
14. Tsimoshynska, O., Koval, M., Kryshtal, H., Filipishyna, L., Arsawan, W. E., Koval, V. (2021). Investing in road construction infrastructure projects under public-private partnership in the form of concession. Naukovyi Visnyk Natsionalnoho Hirnychoho Universytetu, 2, 184-192.

Development of the logistics and transport system of organic products as a perspective direction of Ukraine's environmental strategy

Kateryna Bogomolova[1], *Viktoriia* Kralia[1], *Olha* Podolska[1*], *Yuliia* Sahachko[1], and *Olena* Smihunova[1]

[1]Petro Vasylenko National Technical University of Agriculture, Department of Organization of Production, Business and Management, 44 Artema Str., 61002 Kharkiv, Ukraine

Abstract. The state of development of the organic market in Ukraine is analyzed**.** Shown that organic agriculture is currently positioned in the country as one of the key direction to increase the level of environmental friendliness, economic feasibility in agriculture and is a progressive way to improve the balance between environmental and economic factors of society. The dynamics of changes in the number of organic farms and the area of their agricultural lands is considered. The analysis of the development of the Ukrainian domestic market of consumption of organic products is carried out, and also the dynamics of import of certified organic products from Ukraine to the EU is investigated. Relevance of the issue of logistics by road and rail is proved. The state and main problems of the transport infrastructure of Ukraine are considered. Emphasized that to improve the logistics needs strong institutions, effective coordination of state and government support at the highest level. The development of strategic directions for the improvement of the logistics and transport system is proposed in order to strengthen the competitive position of Ukrainian agricultural producers in the world market of organic products and intensify the integration of the state into the international economic space.

1 Introduction

Today's realities make it clear that the well-being of present and future generations and the level of national environmental security depend on the results of economic entities in all spheres of the national economy. Chosen type of products or services through applied technology they directly affect the environment, as environmental or low-waste technologies will cause much less damage to the environment than intensive and environmentally unsafe.

Thus, the environmental component affects all aspects of the enterprise, namely cooperation with business partners, state environmental control bodies, financial and credit organizations, mutual understanding with the public and as a result on the internal strategy of the organization and management decision-making mechanisms [1]. Clearly, by taking

* Corresponding author: o_podolskaya@ukr.net

into account the opportunities and risks related with environmental problems depends on a successful business.

The practice of modern environmental concept at enterprises in both domestic and foreign markets is not possible without the use of a new format of business. Namely, to integrate all economic subjects involved: producers (goods and services), suppliers of raw materials and components, trade and logistics companies, consumers and other stakeholders into a single environmental cycle. And also to work closely with organizations that promote products on the market, and ensure success in competition, and form the image of an environmentally friendly enterprise. Therefore, the use of environmental management in enterprises is a relatively new phenomenon in the world economy, which can be defined as a special management system aimed at preserving the quality of the environment, ensuring regulatory environmental parameters and based on the concept of sustainable development of society.

Not the best state of the environment, worsening statistics of oncological diseases, force the population to reconsider their attitude to food are daily present at their table. After, understanding the relationship between the state of the environment and product quality, and as a result, human health, leads to a rethinking of food culture and awareness of the value of an ecological product.

2 Features and logistical problems of organic producers in Ukraine

The constantly growing interest in organic food around the world allows Ukrainian farmers to be as full participants in this market, because the available resource potential, technology, provide an opportunity to ensure sustainable competitive advantages. All of the above contributes to the development of ecological or organic agricultural production in our country [2].

Organic agricultural production is an integrated system that combines the best practices in view of environmental protection, maintenance of biological diversity, conservation of natural resources, high standards and production methods. The development of organic agricultural production is a topical and at the same time controversial issue of the functioning of agricultural formations in a competitive environment.

Discussions among scientists and practitioners about the economic feasibility and organizational possibilities of introducing organic production in agricultural formations are related to the conditions of imbalance of the external environment, functioning, increased competition and intensification of the effects of the global environment. The immediate impact exercise existing level and structure of the quality requirements for consumer products, the prospects of integration into the European market, requiring strengthening efforts towards the production of environmentally safe products. At the same time, many years of foreign experience demonstrates the economic, environmental and social benefits of organic production (Fig. 1) [6].

The ecological benefits of organic production are undeniable, it's high-quality food, minimal damage to the environment, ensuring ecological balance at each stage of production, sales of organic products. Spreading of organic production in different countries of the world gives Ukraine the opportunity to find its place among producers and get an additional opportunity to attract investment, financial resources and prospects for innovation.

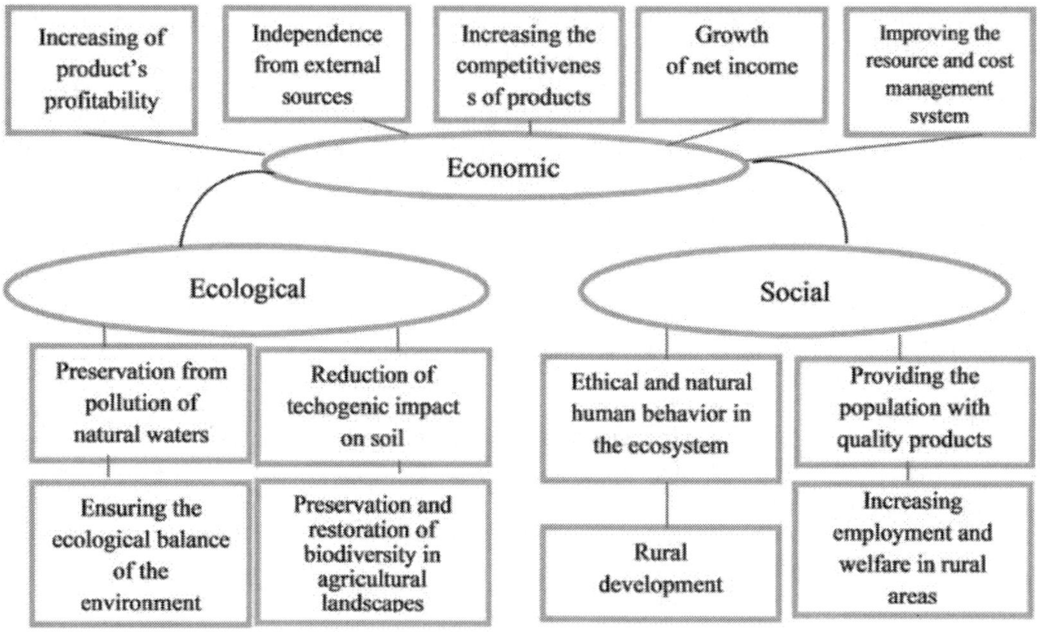

Fig.1. Key benefits of organic production in the world.

Organic products currently in high demand worldwide, and the number of producers is increasing annually. Organic agriculture is tolerant of the environment and is based on the principles of ecology - the science of the environment, and the deep laws of biology - a combination of sciences of natural world. Therefore, it can be argued that organic agriculture is essentially both ecological and biological. Therefore, the use of the terms "ecological" and "biological" does not contradict EU regulatory directives when it comes to organic agriculture. [6, 9].

In 2019 Ukraine took the second place among countries exporting organic products to the EU, behind only China. The part of Ukrainian organic products in the markets of European countries was 10.4%. he European countries that import the most organic products are the Netherlands, Germany, Austria, Poland, Lithuania and others. (Fig.2). These are mainly cereals and oilseeds, oilcake, fruits, vegetables and fruit juices.

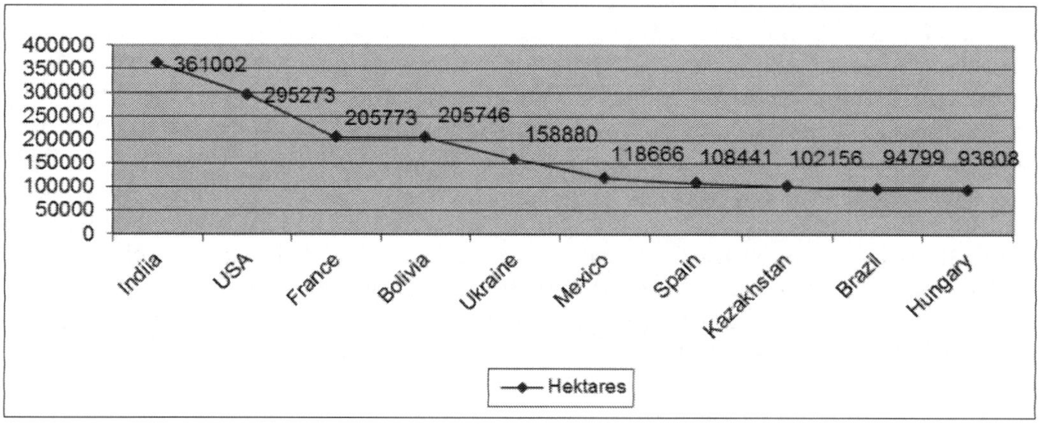

Fig.2. The structure of imports of Ukrainian organic products in the EU.

However, only about 1% of cultivated land in Ukraine is currently occupied by organic farming. Although Ukraine is among the 10 countries with the largest increase in organic agricultural land (Fig. 3).

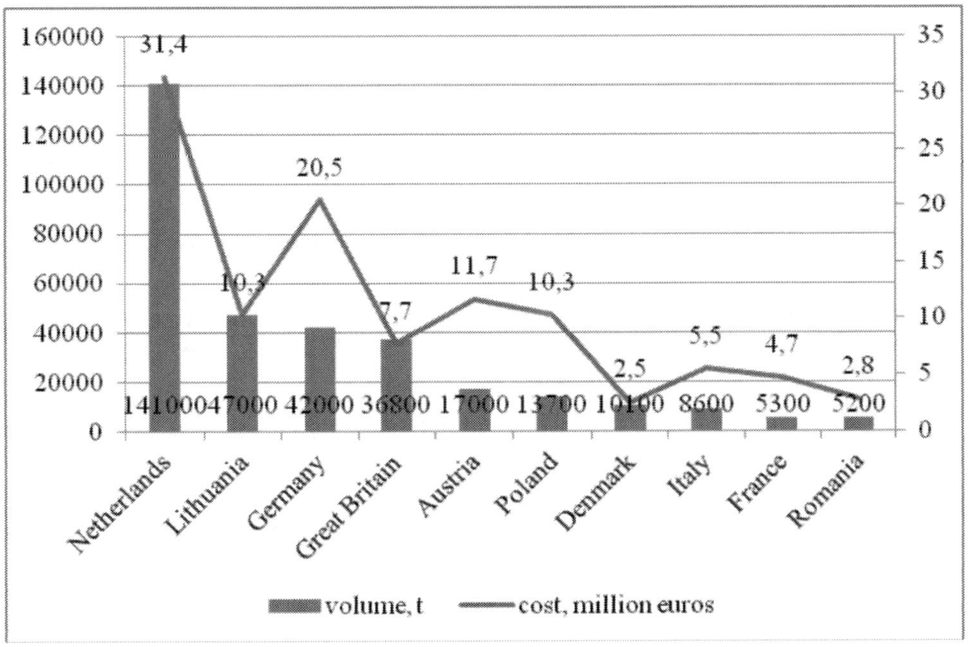

Fig. 3. World: the ten countries with the highest increase of organic agricultural land 2019).

*Source: FiBL survey 2021, based on data from government bodies, the private sector, and certifiers. For detailed data sources see annex (International, Bonn. Available at https://www.organic-world.net/yearbook/yearbook-2019.html).

Factors that constraining Ukrainian farmers fo implementing organic farming technologies are the difficulty of complying with the relevant technological requirements; lower production productivity compared to intensive production technologies, and, as a consequence, the inability to obtain sufficient raw materials capable of ensuring food security of the country, lack of domestic demand for organic products due to low solvency of consumers and so on.

At the same time, analytical data from the commercial service of the US Embassy in Ukraine show that the average return on investment in Ukrainian organic farming is about 300%, which makes it one of the most attractive direction for investment in Ukraine [3]. Often companies in the country have projects on introduction of organic farming with the support of foreign companies (primarily Swiss, Dutch, German) [4].

Promotion of organic agricultural production is also facilitated by public organizations and associations, whose activities are aimed at intensifying the joint efforts of all stakeholders (producers, research and educational institutions, other enterprises, individuals and organizations) in the production and distribution of healthy, environmentally friendly products [9].

Currently, Ukrainian farmers specializing mainly in organic crop production. The total area of agricultural land certified by "Organic Standard" in 2019 amounted to 309,100 thousand hectares, including 233,500 - lands with organic status. At the same time, there were 635 operators of the organic market, of which 501 were agricultural producers [5]. During the cultivation of cereals was 48.1% of land certified as organic, which puts Ukraine in seventh place among producers of organic grain [3].

Organic production involves fertile soils, which without the use of mineral fertilizers are able to provide stable crop yields. The main requirements are strict adherence to crop rotations, application of organic fertilizers and inclusion in the structure of green manure crops to provide nutrients and reproduction of humus in the soil [6].

Today, Ukrainian producers of agricultural products in conditions of fierce competition in the markets have only way to survive is the production and export of organic agricultural products. To export activity by entering into direct contracts today are able to only holdings with strong land bank, excellent reputation and financial independenceAt the same time, small organic producers are practically deprived of such an opportunity due to the influence of the following factors: interest of foreign importing companies in large batches of products, significant transport and overhead costs, which makes the sale of small batches economically unprofitable; absence of developed logistics infrastructure and high logistics risks [11].

Most of farmland are under organic production in southern Ukraine, namely in Odessa and Kherson regions. Today in the Odessa region about 100,000 hectares are certified according to international organic standards. This concentration is due to the natural location of the waterways and the seaport. Therefore, the logistics infrastructure is developed in these areas and therefore plant products are supplied here for storage and further shipment to foreign customers by ship transport.

Some organic agricultural producers from other regions of the land bank of 3,000 hectares or more, which previously shipped products to car trucks are considering the possibility of access to seaports. The only barrier for them is transportation within the country that can increase product price to 20 euros / ton. Therefore, the issue of logistics of road and rail transport is still very topical. Another topical issue is the modernization of terminals in sea and river ports for organic specifics.

3 Strategic directions of solving specified problems

Agri-food sector Ukraine has significant potential for the promotion of organic products both on the domestic and external consumer and industrial markets. One of the ways to achieve this potential is to develop the segment of transport logistics for the promotion of organic products, which is in demand among consumers in the world, and its supply is insufficient in terms of quality content.

The vast majority of Ukrainian organic agricultural products are exported from Ukraine through seaports. Ship deliveries are one of the most economically attractive types of delivery for some countries, and among the advantages is the economic feasibility of forming large batches. In Ukraine there are 33 grain terminals with a maximum capacity 50 million tons Grain per year.

Among the disadvantages of transportation by sea are:

- very long term delivery;
- high risk of pollution or contamination;
- risk of loss or damage due to leaks in holds.

Currently, the transportation of organic products in Ukraine is carried out in the following ways - by rail, water and road (Table 1).

Currently, the factor hindering grain exports is logistical and transport difficulties with grain transshipment. For example, in Ukraine there are currently more than a thousand elevators, which allow power to ship daily to 700 thousand tons of grain. At the same time, Ukrainian Railways wagons are enough to export only 100,000 tons of grain. According to the World Bank, unsatisfactory logistics in the agro-industrial sector is the reason for producers not receiving 0.6-1.6 billion dollars [12].

As for granaries, the vast majority of elevators in Ukraine have a low level of mechanization and need modernization. As of the beginning of 2019, the domestic capacity for simultaneous storage of grain and oilseeds amounted to about 72 million tons with a volume to be stored of 90 million tons, ie the deficit of storage capacity was almost 18 million tons [8].

Table 1. State and problems of transport infrastructure of Ukraine.

Type of transport infrastructure	Advantage	Problems	Ways to solve
Rail transport (about 60% of total grain traffic)	- transportation of grain in containers guarantees quality; - provides convenience during transportation of grain in special containers, - the problem of unloading of highways is solved	- delay in the allocation of grain wagons during periods of active grain exports; - deterioration of Ukrainian Railways production indicators (reduction of average speed, deterioration of grain truck turnover, reduction of cargo transportation due to grain truck surplus).	- clear rules and rates the use of railway infrastructure; - gradual accumulation of private companies' own fleet of hoppers.
River transport (about 3% of total grain transportation)	- a more economically attractive type of grain movement due to minimal costs	- shortage of coasting fleet in Ukraine; - uncontrolled discharge of hydropower water; - the existence of barriers to access to foreign ships in the internal waters of Ukraine.	- extensive work on dredging rivers; - extension of the navigation period in the winter period of the year (November - March); - reforming the navigation system to modern standards, - optimization of gateways and drawbridges; - simplification of permitting procedures and reduction of tariffs for navigation services
Road transport (about 36% of all grain traffic volumes)	- timeliness of product deliveries during periods of active railway loading; - relatively low cost of transportation by road due to fierce competition in the market of transportation services.	- destruction of the road surface and increase of carbon emissions into the atmosphere; - shortage of drivers; - downtime for loading and unloading due to the weak organization of automobile logistics; - violation by agricultural producers of the rules of transportation of organic products; - transshipment of transport by grain; - shadow markets of car services and fuel.	- introduction of services for automation of automobile logistics (for example, Jit + and Zernovoz); - creation of automated points of dimensional and weight control; - reorientation of the market to the production and purchase of equipment for the transportation of grain with a lightweight design.

In order to encourage farmers to build their own granaries Ukraine government developed a program of compensation for 30% of their construction, in that in 2019 allocated 850 million UAH. However, even under such conditions, only agroholding can afford to build their own storage facilities.

To solve the problems of development and improve transport and logistics infrastructure necessary government regulation to encourage competition and create a favorable investment environment and facilitate the creation of additional jobs and value added products [7].

Efficient trade and transport logistics are key to developing the competitiveness of countries around the world. Therefore, it is important to develop strategic directions for the development of the logistics and transport system in Ukraine (Fig. 4.).

Improving logistics services requires strong institutions, effective public sector coordination and the highest level of government support. Transportation of goods often requires the efficient operation of many links in the supply chain. Efficient logistics is also important for the internal market to available, safe and reliable transport and other logistics services could help Ukrainian economy to develop stably own economic, social and environmental aspects [10].

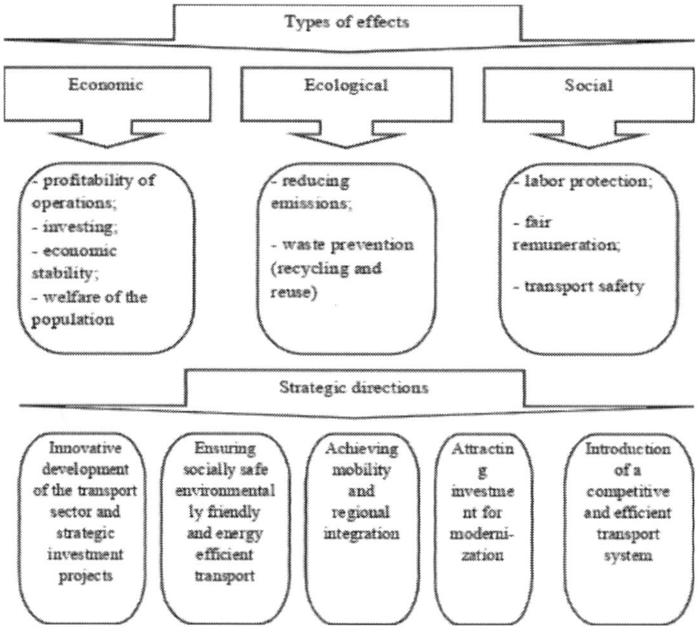

Fig. 4. Strategic directions of development of logistics and transport system in Ukraine.

4 Conclusions

So, with the popularization of healthy lifestyle in the world and in Ukraine relevance of organic products is growing. Ukrainian society is increasingly interested in what is "organic products" or "organic production", which is a consequence of increasing the volume of organic products.

Ukraine, having an agricultural potential, has all the opportunities to fully join the global organic movement with the prospect of becoming a leading supplier of high quality certified products not only to the European market but also to the growing Asian and American markets. In addition, the socio-economic and environmental benefits of organic for Ukrainian society deserve special attention.

Established that, despite significant potential for the production of organic agricultural products, domestic enterprises face a number of obstacles, and especially with imperfect transport logistics. Strategic directions and main ways to address this issue for Ukraine have been developed. This requires a pruden investment policy of stakeholders and government support.

References

1. Shabinskyy, O., Eevdokimova, M., & Furdak, M. (2017). Economic nature and scientific and methodological principles of management of economic security of the agricultural sector. Visnyk of the Kharkiv National Technical University of Agriculture named after P.Vasilenko, 182, 47-59.
2. Bagorka, M.O. (2017). State and problems of development of the market of organic products in Ukraine. Black Sea Economic Studies, (14), 29-33.
3. Agropolit (2018). Analysis of the market of organic products in Ukraine.. https://agropolit.com/spetsproekty/407-analiz-rinku-organichnoyi-produktsiyi-v- Retrived from ukrayini/
4. Wdowith, V.G., & Cherkaska, I.S. (2013). The state of organic production in Ukraine. NUBIP Ukr. Ser. Economics, agricultural management, business, 181(6), 408.
5. Ministry for Development of Economy, Trade and Agriculture of Ukraine (2021). Organic production in Ukraine.. https://agro.me.gov.ua/ua/napryamki/organichne-virobnictvo/organichne-virobnictvo-v-ukrayini
6. Smihunova, O.V., & Ilyin, O.A. (2020) State and prospects of development of organic production in Ukraine. Ukrainian Journal of Applied Economics, 5(4), 98-105.
7. Krylov, Ya.O. (2018). Main directions of grain market development and mechanism of its regulation in Ukraine. Ukrainian Journal of Applied Economics, 3(3), 43-53.
8. Ilchuk, M.M., Konoval, I.A., & Baranovska, O.D. (2019). Development of the grain market in Ukraine and its stabilization. Economika APK, (4), 29-38. http://eapk.org.ua/en/contents/2019/04/29
9. Minenko, S.I., & Dudnik, O.V. (2020). Organic agriculture as a source of increasing the competitiveness of Ukraine in the international arena. Materials of the VII International scientific and practical conference "Actual problems of innovative development and staffing of the agro-industrial complex", June 4-5, 2020, Belarus.
10. Ambarexport (2019). Logistics for grain export. https://ambarexport.ua/blog/grain-export-logistics
11. Goncharenko, N. (2018). Clustering as a tool to increase the competitiveness of organic crop production in conditions of international competition. State and Regions, (4), 106 – 112.
12. Lb.Ua (2016) Grain logistics in Ukraine: barriers to export growth. https://lb.ua/economics/2016/08/15/342521_logistika_zernovih_ukraini.html

Logistics models of modernization of distribution and transmission systems for the development of Ukraine as an exporter of green energy

Natalia Kovtun[1], and *Nadiia* Yushchenko[2*]

[1]Taras Shevchenko National University of Kyiv, Department of Statistics and Demography, 90A, Vasylkivska str., 03022 Kyiv, Ukraine
[2]Khmelnytskyi National University, Department of Automated Systems and Modeling in Economics, 11, Instytuts`ka str., 29016 Khmelnytskyi, Ukraine

Abstract. The article compares scientific facts and analyzes the prospects for the development of the export potential of Ukraine in the field of green energy against the background of an increase in the cost of silver used in the production of solar panels, and due to difficulties in meeting the increased demand for it by mining companies, as well as the expected reduction in the cost of thermonuclear energy due to the rapid implementation of the modified, thanks to advanced artificial intelligence systems, the design of the SPARC reactor. Economic and mathematical models for the regulation of stocks of material resources and equipment, including silver-containing ones, are proposed, which make it possible to study the invariants of managerial decisions and minimize costs. The main factors affecting the volume of formation of stocks are generalized, the types of costs associated with the management of stocks of material resources, the main types of economic and mathematical models for regulating stocks and computer programs available on the market that implement models and methods of supply management and stocks of material resources are systematized.

1 Introduction

So far, a world breakthrough in energy is expected since the creation of the first nuclear reactors, which is associated with the possibility of commercial use of nuclear fusion, which will provide the largest improvement in living standards in history, as well as make virtually every country food and energy independent. Indeed, having large-scale public and private funding, the SPARC reactor project, approved in 2020 at the Massachusetts Institute of Technology as promising to reduce the cost of thermonuclear energy, is rapidly improving thanks to new artificial intelligence technologies and the new technology can be implemented in a matter of years [1].

* Corresponding author: nadezhda15yu@gmail.com

The International Energy Agency (IEA) predicts [2] that by 2025 renewable energy sources will supply one third of the world's electricity, thereby displacing coal from the first place and becoming the largest source of electricity in the world (Fig. 1).

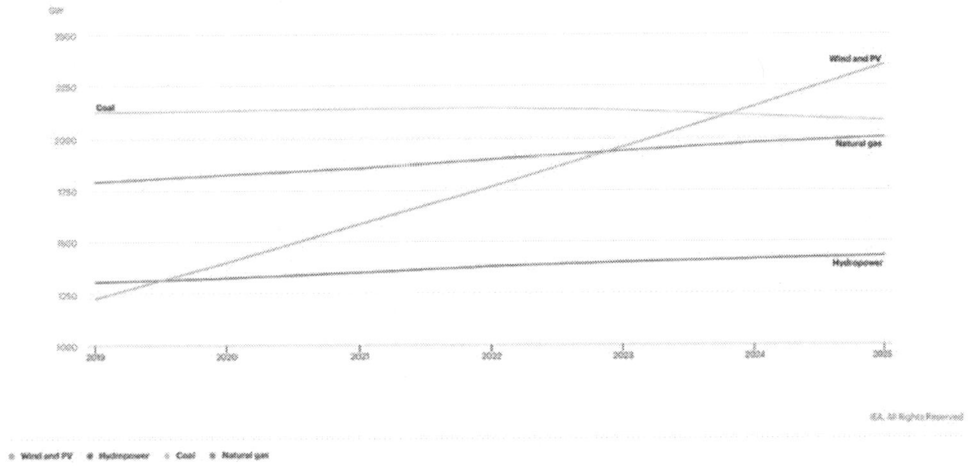

Fig. 1. Total installed power capacity by fuel and technology 2019-2025, main case [2].

Following global trends, realizing that energy efficiency and "green" energy is the key to a competitive economy, energy independence and the health of citizens [3] on sustainable development, "green" transformation, that is, in fact, given its high energy intensity, – on the new economy, and supports the ambitious program of the European Green Deal [4].

In an environment where more and more countries are considering the development of green energy and reducing the impact of the energy sector on the climate as a driver of renewal, a source for economic growth and job creation, from the point of view of the European Union, it may be a good goal for Ukraine to become an exporter of green energy. because the potential is there, the market is created. According to the Minister-Counselor of the European Commission, Thorsten Wollert, "this will be a good new direction and it may attract investments in the future" [5].

According to forecasts of the Danish investment Saxo Bank [6], due to the rapidly growing demand for industrial applications in 2021, there is a real shortage of silver in the market and this undermines political support for investment in solar energy under US President Joe Biden, "European Green Agreement", China's program to achieve carbon neutrality by 2060 and other initiatives.

In order to increase competitive advantages, business entities around the world are rapidly introducing new modern information and communication technologies and solutions for the creation of information resources and the use of digital technologies. The modern software market in the field of logistics is quite large and covers software products of various orientations and characteristics. Material and technical supply systems are being transformed in accordance with modern global trends based on the principles of systematic and logistic approaches, rationality, cost identification, transparency, trusting relationships, equality of approach and ensuring fair competition. There is a shift in the role of supply and material and technical support in enterprise management from solving tactical problems, serving its current needs to solving strategic [7], which increases the urgency of the problem of effective management of stocks of material resources.

2 Methods

Many researchers work in the field of information technology development, methodology, theoretical and practical substantiation of the need for widespread use of adequate economic and mathematical models and methods in managerial decision-making procedures. The issues of the formation of factors influencing the sustainable development of the generating complex of the electric power industry are devoted to the studies of Dontsova O.I., Gibadullin A.A., Kerimova C.V., Morkovkin D.E. [8]. Belyaev N., Chudny V., Egorov A., Korovkin N. [9] are engaged in optimization of the conditions for modernization of heat power enterprises in modern conditions. Mayilyan L.D., Shogenov M.S., Zelentsov L.B. [10] are working on modeling the activities of organizations in the implementation of complex infrastructure projects. However, further formation and development requires economic and mathematical decision support tools for complex systems and capital-intensive projects, in particular, in the energy industry, which allows the subject of management to increase the efficiency of decision-making through modeling and automation of information procedures.

The main goal of this work is to analyze the current state and prospects of modernization of the fuel and energy complex and investments in renewable energy to achieve the Sustainable Development Goals in Ukraine, review, systematization and aspects of the classification of economic and mathematical tools for regulating supplies and inventories necessary for modernizing the energy system and development of Ukraine as an exporter of green energy.

3 Results

The costs associated with the operation of the inventory management system include costs associated with the formation of stocks, costs associated with holding inventory, losses due to lack of inventory or late delivery.

The economic regulation of stocks provides for the determination of a reasonable delivery lot, regulation of the frequency of deliveries, the best terms of shipment and receipt of materials.

In the theory of inventory management, there are the following types of models and corresponding methods for optimizing inventories based on the criterion of minimizing the costs associated with their storage and renewal, which make it possible to reasonably plan the need for material resources, warehouse space and mechanisms, [11, p. 176-184]:

- Wilson's model of determining a batch of an order, subject to uniform demand and the possibility of instant order fulfillment, which minimizes the costs associated with storing and maintaining the stock;
- one-product deterministic static model of optimal inventory management with a possible shortage. The main modifications of the deterministic single-product static model: prohibition of shortages, inability to store stock, replenishment of stocks is carried out instantly at certain intervals, replenishment of stocks is performed instantly, and the occurrence of a shortage is unacceptable;
- dynamic one-product deterministic model of inventory management and production output;
- one-product probabilistic static inventory management model;
- in the case of multi-product stocks, deterministic models are divided into three types: in which the interaction of products is clarified after delivery; with full combination of orders (full completing of batches in accordance with the technological package, etc.); with partial connection of orders.

Numerous invariants of the above models have been developed and proved to be effective.

In practice, the limited resources, as well as the specific conditions of production and delivery, require adjustment of the batch size in comparison with the absolutely optimal, determined, for example, on the basis of the Wilson model [12, p. 214-215]. In particular, the company cannot fulfill large orders due to the limited working capital or the lack of open and closed storage facilities and storage tanks. Conversely, insignificant stocks may turn out to be unprofitable for the supplier due to frequent readjustments of equipment, which reduce its productivity and worsen the use of labor, or because the minimum shipment volume should not be less than the capacity of the container (wagon), since transport tariffs are calculated per container or wagon, regardless of the completeness of the load, and so on.

The economic and mathematical model of the problem of minimizing the costs associated with the formation and storage of stocks, taking into account the restrictions on the storage of several types of stocks, can be as follows.

Determine $q_j, j = \overline{1;\ k}$, ensuring minimum value

$$min: \quad C_T = \frac{R_1}{q_1}\sum_{j=1}^{k} c_{q_j} + \frac{q_1 T}{2R_1}\sum_{j=1}^{k} c_{1_j} R_j, \tag{1}$$

subject to restrictions

$$K = \frac{q_1}{R_1}\sum_{j=1}^{k} c_j R_j \leq K_0, \tag{2}$$

$$f = \frac{q_1}{R_1}\sum_{j=1}^{k} f_j R_j \leq f_0, \tag{3}$$

$$h = \sum_{j=1}^{k} h_j = \sum_{j=1}^{k} \left[\frac{R_j}{q_j}\right] \leq h_0, \tag{4}$$

where q_j – batch size of the j-th resource (products), $j = \overline{1;\ k}$;

C_T – costs associated with the formation and storage of stock to meet the needs of the enterprise in the planning period: $C_T = \sum_{j=1}^{k}\left(\frac{c_{q_j} R_j}{q_j} + \frac{c_{1_j} q_j T}{2}\right)$, but in the case of an assortment of demand, the batch sizes should be proportional to it

$$q_1 : q_2 : \ldots : q_k = R_1 : R_2 : \ldots : R_k$$

or

$$\frac{q_1}{R_1} = \frac{q_2}{R_2} = \ldots = \frac{q_k}{R_k}, \tag{5}$$

therefore, the objective function takes the form (1), that is, it is transformed into a function of one variable q_1;

R_j – demand for the j-th resource in the planning period;

c_{q_j} – costs associated with organizing an order (production) of one batch of the j-th resource (products), regardless of the batch size, monetary units;

c_{1_j} – costs of storing a unit of the j-th resource (products) per unit of time (proportional to the batch size), monetary units;

Т – duration of the planning period, days;
К – variable level of working capital, depending on q, and $К_0$ – a given level of investment, monetary units;
c_j – unit cost of the j-th resource, monetary units;
f_j – warehouse area occupied by the unit of the j-th resource, m^2;
h_j – the number of orders of batches of the j-th resource during the planning period;

(2) - limitation on the amount of available circulating assets, taking into account the range of demand; (3) - the limitation concerning the warehouse area available at the enterprise for storing stocks (taking into account the range of demand); square brackets in (4) represent the integer part of a number.

To determine the minimum of function (1), taking into account the assortment demand, it is necessary to determine the derivative $\frac{\partial C_T}{\partial q_1}$, equate it to zero, then

$$q_1^* = R_1 \sqrt{\frac{2\sum_{j=1}^{k} c_{q_j}}{T\sum_{j=1}^{k} c_{1_j} R_j}} \tag{6}$$

and from the formula (4)

$$q_j = \frac{q_1}{R_1} R_j, \quad j = \overline{1;\ k}.$$

To determine how justified the attraction of additional funds will be by reducing the costs of forming and storing the stock, the conditionally optimal batch size (6) is substituted into inequality (2). If it is fulfilled, solution (6) is final, since it fits into the limited amount of working capital $К_0$. In the opposite case, to determine the extremum of function (1), it is advisable to apply the method of Lagrange multipliers. In this case, an additional function is compiled $F = C_T + (K - K_0)\theta$, where θ is the variable Lagrange multiplier; the partial derivatives of the function F with respect to q_1 and θ are determined, are equated to zero. From the equation $\frac{\partial F}{\partial \theta} = 0$ is determined by q_1; from the equation $\frac{\partial F}{\partial q_1} = 0$ is determined by θ. As a result of simple transformations, q_1 can be expressed through θ. It was reported [13] that with small increases in working capital ΔK_0, the conditionally optimal plan (6) improves, providing a cost reduction $-\Delta C_T$ proportional to ΔK_0 with a proportionality coefficient θ: $\Delta C_T = -\theta \Delta K_0$.

Thus, θ turns into an estimate of the monetary units of current assets: with an increase in circulating assets by 1 monetary unit, costs will be reduced by θmonetary units.

Similarly, on the basis of the Lagrange multiplier method and an auxiliary function $F = C_T + (f - f_0)\theta$, the problem of optimizing reserves with a limitation in areas is solved.

The most essential in order management is the limitation (4) concerning the quantity, and hence the size of orders. Taking into account (4), the objective function will have the form

$$C_T = \sum_{j=1}^{k} \left(c_{q_j} h_j + \frac{c_{1_j} T R_j}{2} \cdot \frac{1}{h_j} \right) \to min. \tag{7}$$

In this case, the enumeration of a number of values h_j satisfying condition (4), and the choice of them conditionally optimal according to criterion (7) is inevitable, since differentiation of function (7) with respect to h_j is inadmissible due to the integer value of h_j.

When taking into account several restrictions at the same time, it should be borne in mind that (2) and (3) limit the lot size from above, and (4) – from below. In the first case, the batch size is selected that satisfies the most stringent condition; in the second, the problem has a solution under the condition of the existence of such batch sizes that satisfy two-sided inequalities.

Optimization of stocks using the methods of mathematical analysis according to the criterion of minimizing the total amount of costs for their storage and renewal makes it possible to reasonably plan the need for material resources, storage sites, storage areas and mechanisms.

4 Discussion

Comparative analysis [14] of software tools available on the market for material flow management by their functionality: procurement / purchasing, warehouse operations, sales and marketing suggests that the most extensive opportunities are provided by BEST (Business Environment Strategic Toolkit), Parus, Galaktika, Axapta packages and Axapta Retail. For use by accountants-analysts, the program "1C: Trade Management" may be appropriate, which has ample opportunities for preparing all the necessary documents, managing the movement of goods and pricing, accepting orders and monitoring their implementation, optimizing warehouse stocks, analyzing goods turnover, planning purchases and deliveries.

The overwhelming majority of software products offered on the market are accounting and require adaptation to specific business conditions. Some have a built-in programming language that allows you to create your own applications. They are expensive and targeted at large corporate structures. However, inventory management problems can be solved by universal means, in particular, MS Excel or MathCAD can be used for optimization problems, for problems using the methods of probability theory and mathematical statistics – Statistica, Gretl and other statistical software products.

References

1. Saxo Group. (2020). Revolutionary fusion design catapults humanity into energy abundance. https://www.home.saxo
2. International Energy Agency. (2020). Renewables 2020. Analysis and forecast to 2025. https://www.iea.org/reports/renewables-2020
3. Cabinet of Ministers of Ukraine. (2021, 29 January). Energy efficiency and green energy are key to a competitive economy and energy independence of communities. https://www.kmu.gov.ua
4. Ministry of Environment Protection and Natural Resources of Ukraine. (2020). Concept of Ukraine's "green" energy transition until 2050. https://mepr.gov.ua/news/34424.html
5. Analytical Center DiXi Group. (2021). Discussion with ambassadors: decarbonization, COP26, green economy. The visit was organized by the British Embassy in cooperation with the DiXi Group Analytical Center. https://dixigroup.org

6. Saxo Group. (2020). Sun shines on silver, which sizzles on solar panel demand. https://www.home.saxo/content/articles/outrageous-predictions/sun-shines-on-silver-which-sizzles-on-solar-panel-demand-08122020
7. Kearney, A. T. (2007). Creating value through strategic supply management. 2004 Assessment off excellence in procurement. https://management-tools.blogspot.com/2007/04/creating-value-through-strategic-supply.html
8. Morkovkin, D.E., Kerimova, C.V., Dontsova, O.I. & Gibadullin, A.A. (2019). The formation of factors affecting the sustainable development of the generating complex of the electric power industry. Journal of Physics: Conference Series, 1399(3).
9. Belyaev, N., Egorov, A., Korovkin, N. & Chudny, V. (2019). Allowance for capacity adequacy criterion in optimizing the prospective structure of electric power system. E3S Web of Conferences, 139.
10. Zelentsov, L.B., Mayilyan, L.D. & Shogenov, M.S. (2019). Organizational and technological simulation of the construction organization activity in the complex infrastructure projects implementation. IOP Conference Series: Materials Science and Engineering, 698(7).
11. Yushchenko, N.L. (2016). Economic and mathematical models in management and economics. Chernihiv : Chernihiv. nat. technologist University.
12. Sakovich, V.A. (1984). Research of operations (deterministic methods and models). Minsk : Higher School.
13. Kulish, S.A., Valovelskaya, S.N. & Rabinovich, I.A. (1974). Mathematical methods in planning material and technical supply. Kyiv : Higher School.
14. Yushchenko, N.L. & Vorokh, M.M. (2016). Comparative analysis of software products that implement models and methods of supply management and inventories. Problems and prospects of economics and management, 1 (5), 74-87.

Suez Canal blockage: an analysis of legal impact, risks and liabilities to the global supply chain

Jade Man-yin Lee[1], and *Eugene Yin-cheung* Wong[1*]

[1]The Hang Seng University of Hong Kong, School of Decision Sciences, 6/F, Lee Quo Wei Academic Building, Hang Shin Link, Siu Lek Yuen, Shatin, Hong Kong

Abstract. The global supply chain currently thrown into chaos due to the Suez Canal incident. On 23 March 2021, the Suez Canal, one of the most heavily used shipping routes, was blocked by vast container ship – the Ever Given. Affecting over 400 vessels scheduled to pass through the Canal in the East-West and West-East directions. Such vessels experienced a dilemma of the use of routes and schedules. Other victims included vessels that were scheduled to arrive at/pass through the Canal, shippers, consignees, ship operators, ship owners, and container terminals. About USD$15 to 17 billion has been held up. This paper aims to reveal the potential damage caused to parties involved and examine the liability of parties involved. Specifically, the impact and legal consequences of the obstruction such as how the scope of general average applies as well as the possible solutions to minimise damage are discussed, followed by future developments. These results suggested that maritime transport support global supply chain linkages and economic growth, any disruption could bring huge impact and costs on various stakeholders along the supply chain. On this basis, the world should learn a lesson from this incident and the need for thoroughness prevention measure should be considered.

1 Introduction

World canals plays an important role in global trade. Construction of canals by developing more direct routes through land provides a shorter routing distance, less time and lower operating costs for vessels. The Suez Canal which is 120 miles long and about 205 metre wide started operating on 17 November 1869 [1]. It is an artificial trading waterway in Egypt, dividing Africa and Asia through the Isthmus of Suez connecting the Mediterranean Sea to the Red Sea [2]. It is a crucial channel for natural gas, transporting cargo and oil. The usage capacity of the Suez Canal in 2020 is massive, amounting to approximately 19,000 ships, or an average of 51.5 ships per day, with a net tonnage of 1.17 billion tonnes [2]. The Suez Canal is an important supply chain channel carrying 13.5% of the World's freight [3]. The Ever Given - the megaship which caused the Incident, is the largest cargo vessel in the world. It is owned by Japanese company - Shoei Kisen Kaisha [4]. The EverGreen, a

* Corresponding author: eugenewong@hsu.edu.hk

Taiwanese shipping company, was the renter and operator of the vessel [4]. The vessel was managed by Bernhard Schulte Shipmanagement and consisted of a full Indian national crew (25 pax) [4]. It is 400 metres long with a width of 59 metres (193 feet) [4]. It can carry a maximum of 20,000 containers at a time [4]. At the time of the Incident, it was carrying 18,300 containers [4].

2. Literature Review

2.1 Canal Blockage Impacts and Prospects

Nowadays, sea transportation is ideal for transporting goods as ocean shipping offers substantial capacity and low prices for international trade. In 2015, more than 80% of international trade was transported by ships [5]. Although the Canal blockage of Suez Canal has been cleared after the incident, it remains an ongoing effect on the global supply chain. More than 300 vessels were waiting to pass through the Canal during the blockage which stopped 12% of world trade from passing through the Canal [6]. The backlog of ships carried everything from cattle to crude oil given that almost USD$15 to 17 billions had been held up [6]. This can be seen from the gas price on the day of the accident which had increased USD$0.40 in the aftermath of the accident [7]. A further wilful effect for the later three to six months is likely [6]. Considering that ships keep coming into the canal, they have to queue at the back of the line given the backlog. Even if traffic has resumed, the backlog has created problems to container availability and ship availability. Countries like the US will experience less impacts than European and Asian countries since the US rely on routes through the Pacific to get goods to the west coast and the Panama canal to the east coast [7]. Though, the west coast routes will be hit by bottlenecks exacerbated by the current pandemic. As such, the US shippers are likely to find alternative routes, for instance air freight, or increase the use of Panama Canal in order to avoid bottlenecks. In contrast, the Suez Canal is the main artery for shipping between Europe and Asia [8]. In particular, the Suez Canal blockage has affected some major development strategies such as the "Belt and Road Initiative" adopted by China [6]. China, as the world's second largest economy, it heavily relies on import and export trade [5]. The Belt and Road Initiative strategy is a huge project with investments in 70 countries and international organisations [9]. It creates a maritime silk road connecting South China with Central Europe and the North Sea [10]. Remarkably, it runs via the Suez Canal. Following the blockage of Suez Canal, this strategy is completely affected by the traffic jam of Suez Canal. Hence, the obstruction shows that one ship can cause enormous damage to the world economy. Expectedly, there will be increase in pricing in the shipping industry especially on air freight pricing as some shippers shifted some of their shipments from sea to the already hectic airline capacity [8]. Since the Suez Canal is one of the five major checkpoints in the world, we may need to reassess the most efficient way to conduct global trade, especially, finding alternative routes.

Alternative routes to substitute the use of the Suez Canal are suggested (Nottebook, 2012). The routes of Suez Canal and Cape of Hope as substitution has considered the distance, transit time and generalised the cost from an origin to a destination for 11 major trade lanes [11]. The results revealed that the Cape of Hope route has the biggest potential to be an alternative to the Suez Canal route. Verny and Grigentin (2009) had evaluated the economic feasibility of the Northern Sea Route, Suez Canal and other multimodal transport alternatives. They observed that the Suez Canal and the Northern Sea lanes provide the least expensive transportation cost of moving goods [12]. Since every shipping company strives to optimise their profit with the lowest transportation cost, the Northern Sea route could also be an alternative route for shipping companies.

2.2 Critics over the Suez Canal

The obstruction brings attention to the importance and effectiveness of the Suez Canal. There were critics over the Suez Canal since it began operating in 1869. The Canal is the key node in the international shipping network. Any blockage of this node will cause network variations to major shipping lanes. Therefore, companies usually focus on the costs, safety and geographical features when making decisions. It allows companies to evaluate the impact of canal comprehensively. The safety of the Canal is a crucial consideration in addressing operation and management of ships as maritime security guarantees ships the ability to sail through the Canal. Lu and Gao (2015) and Lu and Wang (2015) have assessed the safety situation and safety efficiency of the Suez Canal and concluded them poor [13][14]. Rupert found that the location of the Suez Canal presents challenges and vulnerabilities to its practical security [15]. Since it is located within a turbulent and fragile region, it was always targeted by terrorists during wars. Arguably, the monitor security of the Suez Canal is tight. There are multiple Egyptian military units armed with equipment stationed in the Canal zone. It consistently updates the troops and equipment comprehensively. The Canal is also covered by a range of surveillance. The coverage of CCTV is high with cameras situated at each signal station along the canal. Furthermore, the Suez Canal Container Terminal's security works closely with other law enforcement agencies to ensure the security of the terminal also applies to the nearby rivers and paths to the northern end. The way to access the terminal is also heavily guarded where positive identification must be present.

All shipping companies want to improve their transport efficiency and reduce cost. In the maritime industry, transportation costs are significant to shipping managers. Thus, companies are looking for the shortest route with the lowest cost. Clearly, longer shipping distances will increase the cost. However, it is not the sole factor influencing the costs. Factors such as cargo categories, ship type, sailing speed also need to be taken into account. Notably, the Suez Canal is known as one of the most reasonable shipping routes [5]. As cost is one of the most significant factors affecting the decision making of businesses, its lower cost compared to other routes is very ideal. However, many issues arise when the Suez Canal is blocked. If this is the case, the shipping routes will be longer likewise the transport cycle will also become lengthy. Many vessels have to abandon their original shipping route and deviate to the closest but lengthier route - the Cape of Hope, in order to avoid further delays. As a result, distances were increased by between 4000 and 6000 nautical miles and transportation expenditure increased around 4 to 5% [5]. The blockage will also cause inflation to the container transportation cost. Considering Europe takes up the highest proportion of containers in world shipping trade at 23%, the blockage will induce more than 10% increase in cost percentage per container [5]. Therefore, the greater increase in transportation costs reveals a bigger impact of the blockage.

In addition, there are several incidents at the Suez Canal over the years due to the dimensions of the vessels. The Suez Canal's length is about 193.3 kilometres with a depth of 28 metres and its widest point located at the Great Bitter Lake is only 133 metres wide [17]. The Suez Canal Authority (SCA) has set out limitations on a vessel's width and depth, however, there were no restrictions on the length of vessels. Instead, vessels over 400 metres long must getting permission from the SCA before navigating the Canal. Yet, many largest cargo vessels in service are 400 metres long [18] like the Ever Given which was stranded in the Suez Canal in 2021, has a Suezmax size of 399.9 metres long [19] that just fits within the limit set for the Canal. With the capacity to carry 20,000 containers, the Ever Given is unlikely to stop suddenly. So, when operating the Suezmax within the narrow margins, the skipper has to maintain the speed steadily in order to keep the vessel steering effectively. The process to salvage and refloat such vessels is complicated. It requires specialised instruments and could be time-consuming depending on the severity of the grounding [20]. Hence, this makes a blockage like this almost unavoidable, particularly because the ship's length far exceeds the canal's width.

2.3 Legal aspect

Apart from the operation and management of the ship, the legal rights and potential legal disputes on the canal blockage are evaluated as intensive lawsuits and challenges would be encountered by various stakeholders along the supply chain. The principles governing the shipping industry is dictated by the law of the sea or maritime law. The former law applies internationally while the latter refers to domestic shipping law. These laws play a crucial role in solving disputes between parties.

2.3.1 The implication of a contract of affreightment

A contract of affreightment refers to an agreement allowing the cargo owner to enforce the shipowner himself or through an agent to transport goods by sea. Such agreements may take various forms. The most common forms are the charterparty and bill of lading. A charterparty is a contractual agreement between the shipowner and the charterer for the use of vessel entirely or partially [21]. A billing of lading is an agreement between the carrier and the shipper of goods [21]. It outlines the details of goods to be shipped including type of goods, destination and quantity etc.

All contract of affreightment has a mandatory framework, the Hague-Visby Rules, that set out minimum responsibilities, liabilities, rights and immunities of carriers. Outside the Hague Visby Rules, parties are free to negotiate additional terms under the principle of freedom of contract. The additional terms could be expressed or implied. In the course of business, terms are usually expressly written in the contract. Sometimes, such terms might strike down the minimum liability as provided under the Hague Visby Rules. For instance, a bill of lading incorporates the Hague Visby Rules illustrating that the vessel cannot deviate from voyage. However, parties included a term in their contract stating that "Where a carrier is affected by grounding, he therefore will not be liable for deviate from the usual route.". In such case, the parties' term prevails the Hague Visby Rules and the carrier is not in breach of contract.

Noticed that some vessels navigated to alternative routes in order to meet deadlines following the Suez Canal blockage incident. However, it might have an unexpected consequence. In Société Franco Tunisienne d'Armement v. Sidermar S.P.A [22], the shipowner chartered the vessel to the charterparty for the purpose to sail goods via Suez Canal. Later, the suez canal announced closure of the canal. The shipowner asserted that the vessel can navigate via alternative route, the Cape of Hope which can also reach the same destination and wanted to increase freight charge while the charterparty claimed that the contract is frustrated. The court concluded that the contract was frustrated and the alternative route is circuitous, unnatural and different in a variety of ways from the original route. Applying to the Suez Canal blockage, if the alternative routes navigated is too absurd from the original purpose and routes, it might cause frustration to the contract. Frustration to contract means the vessel operator is liable for the breach of contract. The cargo owners will then be able to sue the vessel operator for absurd deviation and the operator will not be able to restitute from the extra transportation expenditures.

2.3.2 The application of laws

As a contract of affreightment relating to international sales of goods, this inevitably raises issues to the applicable law over the contract since rights and obligations are different in different legal systems. Although maritime law has developed internationally, a contract usually operates under the laws of an individual country. Remarkably, outside of the Hague-Visby Rules, parties can freely negotiate additional terms following their wishes. Hence, most contracts will contain an express term regarding to the choice of law. This practice has long been customary for parties entering into contracts of affreightment to prevent any argument relating to the choice of law which governs the contract. For instance, a contract is about to co-signed by one party located at Netherland while another in China. Before the contract is formed and signed, it is essential for them to demonstrate which country's law they will use in order for them to incorporate rules of that country into the contract.

2.3.3 The role of maritime insurance

Today, parties usually collaborate with maritime insurance in settling claims hence it plays an essential role in the shipping industry. It is a contract between the insured and the insurer by means of protection and indemnity against a consideration to be paid by the insured in the form of a specific premium. The insurance is applicable to the perils of sea as well as risks relating to cargo, ship and freight such as loss of cargo or cargo damage in which goods are transferred between an origin and a final destination. This guarantees the safety of the ship and the safe arrival of goods.

Maritime insurance is classified into 3 categories, namely, Hull insurance, Machinery insurance and Protection and Indemnity (P&I) insurance. The Hull insurance covers damage to the vessel while the Machinery insurance compensates all essential machinery of the ship [23]. Lastly, the P&I insurance protects risks connected with ownership of the vessel and risks related to the hiring of the ship such as crew related claims and cargo related claims [23]. P&I insurance is a mutual insurance association providing shipowners a risk pool through mutual maritime insurance. It covers the liabilities to third parties and risks which are not covered elsewhere in the hull and machinery insurance and other policies [23].

2.3.4 Unique insurance coverage to maritime industry - General Average

General average is an example of P&I insurance. It is a principle peculiar to maritime law. It is an international regulation, and it aims to preserve a vessel or a life from a dangerous situation in an emergency. Hence, a vessel or life must be saved in order to declare a general average. It compels parties involved to share any losses occurred from voluntary sacrifice in the event of an emergency. Damage to the hull while attempting to refloat is an example of sacrifice. In contrast, if the entire vessel is lost then it cannot be declared as a general average.

Essentially, general average has nothing to do with contract of affreightment. However, in the 14-15th century, the invention of contract propose the general average to be implemented into the contract [24]. Thus, in the late 19th century, the York-Antwerp Rules were adopted to uniform the principle of general average [24]. The York-Antwerp Rules have set out what losses and expenses are admissible in general average and how general average is to be apportioned between parties [24]. Therefore, in most cases today, general average is usually incorporated into charterparties or bills of lading. Hence, whether a party is liable to another will subject to contractual clauses.

A general average expenditure includes the cost of discharging cargo to refloat the vessel and hiring of tug to assist with the refloating of vessel. To define the general average contribution, the shipowner needs to appoint a general average adjuster to determine the cost incurred by each involved party.

A general average adjuster is an expert on maritime insurance and general average. He or she has the duty to confirm the general average amount. In order to calculate the total, he or she has the duty to examine documents such as assess cargo manifest, bill of lading and deposit receipts in order to find out what is allowed under the general average, contributory values and allowances. The shipowner may discharge the general average sacrifice by cash deposit, signing of Lloyd's Average Bond backed by bank guarantee or through its P&I club or similar institution guarantees. This means in cases where the shipowner sustains losses due to a sacrifice e.g. delay of vessel, all parties who were concerned with the safe arrival of the ship such as the charterer, shipowner, cargo owner will contribute to these losses. The contribution may be covered by institutes like the P&I club which is a non-profit co-operative formed by shipowners where each member contributes a certain amount to a common risk pool. Therefore, such losses could be covered fully or partially by the P&I club.

2.3.5 Latest judgment of other accident in the Suez Canal

There was a collision incident in the Suez Canal on 15 July 2018 [25]. Suffered from engine failure, the vessel at the head of a convoy, AENEAS was halted in the canal [25]. Contingency plan to moor the vessels in the queue behind was taken. During the mooring operation, PANAMAX ALEXANDER collided with SAKIZAYA KALON [25]. Subsequently, these two vessels collided with OSIOS DAVID [25]. The issue arises on who is liable to the collision. On 5 October 2020, the case finally presented to the Court. The Court held that the PANAMAX ALEXANDER provided opportunity for later collision therefore is responsible and liable for the causation. Whilst there was nothing SAKIZAYA KALON and OSIOS DAVID could do to avoid the incident hence they are not liable to the collision [26].

Overall, it will depend upon the investigation outcome and takes ages to solve the entire issue. Whether the insured could claim against its insurer successfully will depend upon the policy features. The incident must also happen within the policy period.

3. Suez Canal Incident

At around 07:40 EGY (05:40 UTC) on 23 March 2021, the Ever Given was enroute from Malaysia to the Netherlands (Fig. 1) [2]. Weather forecasts revealed that the winds on that day were strong where they reached 40 kn (74 km/h; 46 mph) [2]. While the Ever Given was crossing the southern end of the canal, the force of the strong wind caused the head of the vessel to deviate, leading it off course and being stuck on the southern canal [27].

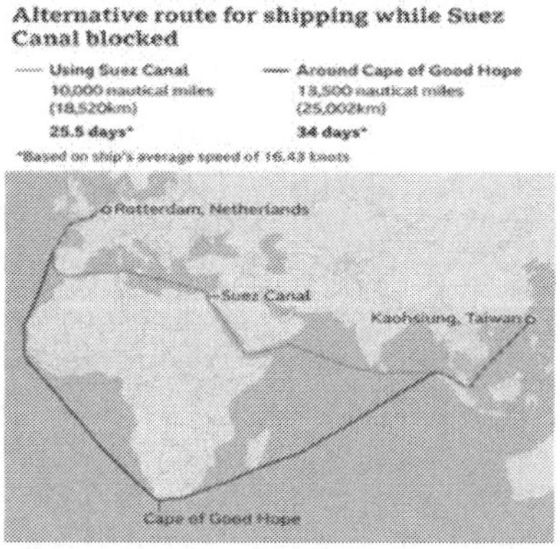

Fig. 1. Alternative routings for ship liners during Suez Canal Blockage.

3.1 Recovery Action

In the aftermath of the Incident, the SCA's collaborated with Dutch Company - Royal Boskalis Westminster's subsidiary - SMIT Salvage to carry out marine rescue operations immediately (Fig. 2) [28]. In the beginning, the owner of Ever Given cooperated with Lloyd's Open Form in an attempt to refloat the vessel [2]. Tugboats were used to tow Ever Given together with an excavator to remove sediments beneath the bow of the Ever Given (Fig. 3) [29].

Fig. 2. Marine rescue operations on Ever Given [4, 5].

Fig. 3. Tugboats towing Ever Given.

Meanwhile, some affected ships took a 3,000 miles detour to Africa's Southern tip - the Cape of Good Hope [30]. The alternative route is lengthy. The journey added more fuel costs and took additional two to three weeks to complete [30].

On 28 March 2021, Ever Given made an announcement saying they were working on the dredging operation and more than 20,000 tons of sand and mud had been removed successfully, loosening the vessel's bow within the bank of the Canal [31]. Also, the engineer had taken advantage of high tide to make Ever Given refloat [31].

On 29 March 2021, Ever Given was finally freed. Engineers took advantage of high tide plus the pulling power of 14 tugboats to refloat Ever Given eventually [28]. To examine the cause of the Incident, Ever Given had begun towing north to the Greater Bitter Lake for further investigation. Soon afterwards, the maritime traffic resumed and 450 ships gradually and successfully passed through the canal [28]. Meanwhile, approximately 400 ships were still awaiting to pass through the Canal in the Red Sea, Mediterranean Sea and the Great Bitter Lakes [28].

On 1 April 2021, the owner of Ever Given declared a general average as a result of the refloating work, which means that shippers would be expected to share the costs of the ship's rescue [32].

On 3 April 2021, the backlog of ships postponed by the disruption was finally cleared [32].

On 12 April 2021, the Egyptian Government demanded $1 billion in compensation to have the Ever Given released from the Great Bitter Lake. The Ever Given refused and the vessel is still being held up by the Egytian government. Despite that Ever Given refloated, the blockage effect over the global supply chain continues [33].

3.2 Impact

Available logistics is one of the presumptive designs of the global supply chain with the focus on price negotiation with an assumption on availability and reliability. Since global trade was interrupted by Ever Given, global supply chains are heavily affected and will remain blocked until the Ever Given gets back on track. The impacts include impacts on logistics operations, legal and insurance with parties not limiting to the carrier, shipper, terminals, charter party and insurers.

Before the Incident, the COVID-19 pandemic had caused a serious setback for the maritime industry. The cost of shipping has doubled since the virus outbreak. The Suez Canal Incident is putting greater pressure on global trade. The blockage leading to a huge congestion of over 370 ships at the Suez Canal. Vessels of all types were forced to wait for the obstruction to clear in order to pass through the Canal. The blockage had caused disruption to deliver over $9 billion worth of goods each day which is equivalent to $400 million worth of trade per hour [34]. Since it had been stuck for over 6 days, approximate $54 billion trade loss had been estimated [35].

In conjunction with trade loss, the actual economic damage around the world could be even higher considering essential commodities including oil and gas may be in interim shortage as a result of the blockage. According to Allianz, a German insurer, the obstruction could diminish global trade growth by 0.2 to 0.4 annually [35]. Subsequently, prices for some commodities or goods may be inflated. Companies may even incur additional expenditure such as shipping delays compensation and rising commodity prices.

According to Fitch Ratings, the insurance sector will also suffer over approximately USD $31 billions loss since insurers are liable to compensate victims, namely companies, customers, other vessels companies etc [36]. Hence, insurers will face a great number of claims. Companies and law firms are also expecting to bring extensive lawsuits regarding cargo interests to courts. Such claims would likely focus on the delay and disruption caused by Ever Given. The damages that can be claimed will depend on the economic loss, damage to freighter and additional expenses incurred. However, it is hard to draw a line on the damage suffered. Thus, the issue would not be solved until all the cases are settled. As a result, the impact on the global supply chain may last longer than months, potentially for years.

As regards to political impacts, the blockage has stopped many countries' plans since Suez Canal is a major shipping lane connecting Europe and Asia. This brings about significant hurdles to global infrastructure development, "One Belt One Road" adopted by China [6]. In this globalised world, many items like mobile phone are completed with components from multiple different countries. The delay caused by blockage will affect the manufacturing industry as well. Since the delay will affect components to be delivered on time, potentially, it might cause components shortage, eventually delaying factories to compose components together. Thus, there is going to be a challenge to get customers what they want.

Although Ever Given was freed, it is believed that the domino effect of the Suez Canal blockage is vast and will last for several years.

3.3 Further Investigation and Possible Outcomes

In order to discover the causes of the Incident, investigations of seaworthiness and Ever Given Captain's action would be conducted. Hence, Ever Given will remain in the Great Bitter Lakes area till the inquiries are over. it is expected that the inquiries will last for a maximum of four days. Egypt has conducted its own examination including interviewing the captain and crew and reviewing the data record in the "black box" of Ever Given [37]. As the registry of Ever Given, Panama's Maritime Authority also carried out its own probe [38]. If the investigation results are contradictory, the cases will become much more complicated.

Some believe that the weather was the cause of the Incident though the reason may not be that straightforward, given that the ships behind Ever Given did not run into similar troubles as the propulsion engines with rudders run by ships are nearly identical and similar. Software and human errors may be variable factors. As noted by the manager, Bernhard Schulte Shipmanagement that was responsible for the crew, thus, if this is the case, the operator, EverGreen would be able to sue it for failure to manage Ever Given reasonably.

As noted that the Ever Given has been grounded by the SCA for fining, there are battles between the shipowner insurer and the SCA. The SCA blamed the Ever Given Captain for sailing too fast while the UK club argued that the convoy was controlled by the SCA pilots and the SCA vessel traffic management services including control of ship's sailing speed and the availability of escort tugs [39]. The case is currently awaiting for the final verdict on compensation which will be set on 20 June 2021 [39].

3.4 Potential Claims and Losses

The shipowner is liable for the damage caused as well as damage caused to third parties. The UK P&I Club is the insurer of the owner of Ever Given. The insurance covers certain third party liabilities, shipowning expenses and expenses arising from operating ships. Considering the Incident, the P&I Club has to pay damages for damage caused to the Suez Canal. Additionally, it might need to recover the shipowner from the paid declared general average after the grounding of Ever Given. It is also liable to compensate victims (e.g. companies and other canal users) who suffer from cargo damage, loss of market, business interruption. It also has to compensate the loss of revenue of the SCA and possible fine from the SCA.

As a renter and operator of Ever Given, EverGreen is responsible for all the goods on board. Since it has the Charterer's Liability Insurance, its insurer, TT Club is responsible for reimbursing the general average and salvage to all the partners involved with EverGreen. This covers all containers and fuels on board. Whether the partners could claim or not will depend on the contractual liabilities. Besides, EverGreen may have to deal with potential consortium claims. Consequently, EverGreen itself will undergo loss of goodwill.

During the investigation, there is an ongoing negotiation between the SCA and the owner, insurer and operators of Ever Given. On 13 April 2021, the Court ruled that USD$916 million damages were to be awarded to the SCA and granted it a warrant to seize Ever Given following the unsuccessful negotiation between them [33]. The SCA emphasised that it will hold up the ship until roughly USD$1 million damages are paid [33]. Meanwhile, the goods on board are stranded and cargo owners are facing additional expense, delay and customer complaints. Further challenges are faced by the operator of Ever Given. It might need to unload the 18,000 goods-filled containers to other vessels [33].

Maritime Law indicates that in huge casualty events, cargo owners on board a ship should contribute to the rescuing cost. The 2018 fire on the Maersk Honam is an example where the cargo owners were subjected to pay off fixed sums amounting to 54% of cargo value in order to release their cargo [40]. Applying to this case, this means that the container owners on board Ever Given may need to pay an additional cost to have their cargo released.

Bernhard Schulte Shipmanagement, the technical manager of Ever Given was in charge of its operation [33]. It has the obligation to ensure the crew handle Ever Given without fault. In case that investigation finds that that the crew was in fault, the operator will have the right to claim damages from Bernhard Schulte Shipmanagement. Except if the Bernhard has purchased the professional Indemnity Insurance, it would be settled by the insurer.

Other affected charterers may want to seek indemnity for cargo damage and late delivery charge from the shipowner. Notably, in place where Hague-Visby Rules apply, it would be unsuccessful to claim for pure economic loss. However, this could be exempted by contractual terms and other rights may be granted. Therefore, whether such claims are allowed or not will be based upon the terms and place the contract was signed.

3.5 Solutions

There is a history of disruption to the Suez Canal. In 2004, the Suez Canal shut down for 3 days due to an oil tanker ship, MT Tropic Brilliance got lodged sideways [41]. Another disruption was found in 2007 where the OOCL Japan vessel was stuck perpendicularly causing the Suez Canal to be closed for a few hours [41]. More incidents were found in the year of 2018 and 2020. Learning from these unfortunate blips, parties could minimise the damage by considering the following.

The above revealed that the Egypt Government is compelled to consider further development of the Suez Canal. Considering that the channel of the Suez Canal is not wide enough for big ships to pass through like the Ever Given, it might want to widen the Suez Canal channel. The SCA should implement a set of contingency plans along with regular manoeuvre. It may interconnect with other countries in such events as well. It should also review ways vessels travel through the Suez Canal by implementing maritime traffic controls such as setting limits on vessel speed, length, weight and height or number of vessels pass through the canal at the same time.

As to the owner and operator of Ever Given, they should send engineers to scrutinise the vessel regularly and before it sails in order to ensure the vessel is good to go. To further protect themselves from legal action, they should ensure their insurances are well covered for such huge casualty events.

Since the Suez Canal incident revealed how fragile the channel between Europe and Asia is and prompted international effort to discover alternative maritime routes. It gives charterers and containers users a good opportunity to explore other routes or forms of transportation as a backup plan. The use of the North Sea Corridor could be an alternative scheme. The "International North-South Transport Corridor (INSTC)" project has been created by Russia, Iran and India since 2002 [42]. Recently, the member states are planning to expand the seaway to a 7,200km long route connecting Mumbai to Moscow [43]. This will accelerate trade activities between Europe and Asia. It is likely to reduce travel times by 20 days and save at least 30 percent shipping costs compared to the Suez Canal [44]. Apparently, the Northern Sea Route has the potential to become another primary chokepoint for the Maritime industry. Nevertheless, whether it could be formed will depend on the foreign affairs between the member states. The future development of this is one to pay attention to.

4 Conclusion

Undoubtedly, the Suez Canal Incident has disrupted the global supply chain. Exposed to such disruption to the global supply chain and experienced in surging prices on goods like oil and gas due to the incident, it demonstrated the need to review the way global trade is being conducted. In case such an event reoccurs, contingency plans should be implemented interstate. To minimise loss, companies should take immediate action to resolve both short term and long-term issues. Companies may consider the use of other chokepoints like the Panama Canal and Northern Sea Route. Evidently, incidents like the Suez Canal are unavoidable, companies should therefore reflect upon the crisis and raise their awareness for their future development.

Reference

1. Suez Canal, Egypt. (2021). Earth Watching. https://earth.esa.int/web/earth-watching/image-of-the-week/content/-/article/suez-canal-egypt
2. BBC News. (2021, March 24). Egypt's Suez Canal blocked by huge container ship. https://www.bbc.com/news/world-middle-east-56505413
3. Cecere, L. (2021, March 29). The Ever Given Is Moving But Your Supply Chain Will Not. Forbes. https://www.forbes.com/sites/loracecere/2021/03/29/the-ever-given-is-moving-but-your-supply-chain-will-not/?sh=1cef5b92744f
4. BBC News. (2021, March 25). Suez Canal: Owner of cargo ship blocking waterway apologises. https://www.bbc.com/news/world-middle-east-56522178
5. Gao, T., & Lu, J. (2019). The impacts of strait and canal blockages on the transportation costs of the Chinese fleet in the shipping network. Maritime Policy & Management, 46(6), 669–686. https://doi.org/10.1080/03088839.2019.1594423
6. Duerr, G. M. E. (2021,March 30). Suez Canal Blockage: Ship Freed, Heavy Traffic, International Impact. DigitalCommons@Cedarville University. https://digitalcommons.cedarville.edu/history_and_government_media_contributions/1231
7. LeBlanc, J. (2021). Suez Canal Blockage: Ripple Effect on Miami Valley Supply Chain.https://digitalcommons.cedarville.edu/business_administration_media_contributions/120
8. Kickham, V. (2021, April 1). Supply chains brace for ripple effects of Suez Canal blockage. DC Velocity. https://www.dcvelocity.com/articles/50093-supply-chains-brace-for-ripple-effects-of-suez-canal-blockage
9. Week In China. (2016, November 1). What is Belt and Road? https://www.weekinchina.com/chapter/belt-and-road/what-is-belt-and-road/
10. Belt and Road Initiative. (n.d.). Belt and Road Initiative. Retrieved May 3, 2021, from https://www.beltroad-initiative.com/belt-and-road/
11. Notteboom, T. E. (2012). Towards a new intermediate hub region in container shipping? Relay and interlining via the Cape route vs. the Suez route. Journal of Transport Geography, 22(2), 164–178. https://doi.org/10.1016/j.jtrangeo.2012.01.003
12. Verny, J., & Grigentin, C. (2009). Container shipping on the Northern Sea Route. International Journal of Production Economics, 122(1), 107–117. https://doi.org/10.1016/j.ijpe.2009.03.018

13. Jing, L., & Wang, S. (2015). Safety Evaluation of China's Maritime Transport Key Nodes. Journal of Transportation Systems Engineering and Information Technology, 15(1), 30–36. http://en.cnki.com.cn/Article_en/CJFDTotal-YSXT201501007.htm
14. Jing, L., & Gao, T. H. (2015). Efficiency of safety control in key nodes of international sea lanes. China Soft Science, 10(1), 1–8. https://doi.org/10.3969/j.issn.1002-9753.2015.10.001
15. Rupert, H. (2009). The Suez Canal: Strategic & Operational Security Realities - Past, Present, & Future. Strategic Insights: Global Maritime Analysis, 19(1), 1–6. https://www.sldinfo.com/wp-content/uploads/2009/11/Securing-the-Suez-Canal-October-20091.pdf
16. Canal Characteristics. (n.d.). Suez Canal Authority. Retrieved May 6, 2021, from https://www.suezcanal.gov.eg/English/About/SuezCanal/Pages/CanalCharacteristics.aspx
17. Park, N. K., & Suh, S. C. (2019). Tendency toward Mega Containerships and the Constraints of Container Terminals. Journal of Marine Science and Engineering, 7(5), 131. https://doi.org/10.3390/jmse7050131
18. Vessels Details. (n.d.). American Bureau of Shipping. Retrieved May 6, 2021, from https://www.eagle.org/portal/#/absrecord/details
19. Hopcraft, R. (2021, March 26). Commentary: Why the Suez Canal accident is a worst-case scenario for global trade. CNA. https://www.channelnewsasia.com/news/commentary/suez-canal-container-ship-accident-global-trade-14486954
20. Baughen, S. (2012). Shipping Law (5th ed.). Routledge.
21. Société Franco Tunisienne d'Armement v. Sidermar S.P.A [1960] 3 W.L.R. 701
22. Gürses, Ö. (2017). Marine Insurance Law (2nd ed.). Routledge. 1-460
23. Wilson, J. (2010). Carriage Of Goods By Sea (7th edition). Longman. 1-559[42] Baghirov, O. (2021, April 18). Suez Canal Shutdown revealed the importance of the Middle Corridor. Modern Diplomacy. https://moderndiplomacy.eu/2021/04/18/suez-canal-shutdown-revealed-the-importance-of-the-middle-corridor/
24. Owners of the Vessel Sakizaya Kalon v Owners of the Vessel Panamax Alexander. (2020). National University of Singapore. https://cmlcmidatabase.org/owners-vessel-sakizaya-kalon-v-owners-vessel-panamax-alexander
25. Sakizaya Kalon (2020). Owners of The Vessel v Panamax Alexander, Owners of The Vessel [2020] EWHC 2604 (Admlty)
26. Farrer, M., & Safi, M. (2021, March 24). Giant ship blocking Suez canal partially refloated. The Guardian. https://www.theguardian.com/world/2021/mar/24/huge-container-ship-blocks-suez-canal-evergreen
27. Schuttevâer, M. L. D. (2021, March 29). Suez Canal unblocked: "We pulled it off!" GlobeNewswire. https://www.globenewswire.com/news-release/2021/03/29/2200887/0/en/Suez-Canal-unblocked-We-pulled-it-off.html
28. Check, R., & Journalism, V. (2021, March 29). Suez Canal: How did they move the Ever Given? BBC News. https://www.bbc.com/news/56523659
29. Talmazan, Y. (2021, March 30). Suez Canal is open again, but impact of Ever Given saga will be felt for some time. NBC News. https://www.nbcnews.com/news/world/suez-canal-open-again-impact-ever-given-blockage-will-be-n1262453
30. BBC News. (2021, March 28). Suez Canal: Effort to refloat wedged container ship continues. https://www.bbc.com/news/world-middle-east-56547383
31. Leonard, M. (2021, April 8). Ever Given, general average and why shippers will share the costs of a ship's rescue. Supply Chain Dive. https://www.supplychaindive.com

32. Schuler, M. (2021, April 14). Ever Given Updates from the Ship Manager – Ship Arrest 'Extremely Disappointing.' GCaptain. https://gcaptain.com/ever-given-grounding-the-latest-from-the-ship-manager/
33. Dzhanova, Y. (2021, March 29). The Suez Canal has a contentious history and has been blocked and closed several times since opening. Insider. https://www.businessinsider.com
34. Das, K. (2021, March 30). Explained: How much did Suez Canal blockage cost world trade. India Today. https://www.indiatoday.in/business/story/explained-how-much-did-suez-canal-blockage-cost-world-trade-1785062-2021-03-30
35. Suez Canal Blockage a Large Loss Event for Global Reinsurers. (2021, March 29). Fitch Ratings. https://www.fitchratings.com/research/insurance/suez-canal-blockage-large-loss-event-for-global-reinsurers-29-03-2021
36. Egypt's Suez Canal Authority reveals how MV EVER GIVEN went aground, holds shipmaster responsible for the accident. (2021, May 31). EgyptToday. https://www.egypttoday.com/Article/1/104488/Egypt%E2%80%99s-Suez-Canal-Authority-reveals-how-MV-EVER-GIVEN-went
37. Egyptian Investigators Kick Off Probe in Suez Canal Blockage. (2021, March 31). GCaptain. https://gcaptain.com/egyptian-investigators-kick-off-probe-in-suez-canal-blockage/
38. Chambers, S. (2021, June 4). Suez Canal Authority to blame for Ever Given's high speed: UK P&I Club. Splash247. https://splash247.com/suez-canal-authority-to-blame-for-ever-givens-high-speed-uk-pi-club/
39. Hochfelder, B. (2018, May 22). Shippers will foot more than half the bill to recover cargo after Maersk fire. Supply Chain Dive. https://www.supplychaindive.com/news/maersk-honam-fire-cargo-risk-shipper-payment/524076/
40. koh, A., Wardany, S. E., & Clark, A. (2021, March 24). Suez Canal Snarled by Giant Ship Choking Key Trade Route. Bloomberg. https://www.bloomberg.com/news/articles/2021-03-23/suez-canal-traffic-blocked-by-container-ship-stuck-in-waterway
41. Sahakyan, M. (2020, September 17). Rebuilding Interconnections: Russia, India and the International North-South Transport Corridor. AsiaGlobal Online. https://www.asiaglobalonline.hku.hk/rebuilding-interconnections-russia-india-and-international-north-south-transport-corridor
42. Baghirov, O. (2021, April 18). Suez Canal Shutdown revealed the importance of the Middle Corridor. Modern Diplomacy. https://moderndiplomacy.eu/2021/04/18/suez-canal-shutdown-revealed-the-importance-of-the-middle-corridor/
43. Business Insider India. (2021, April 5). Suez Canal issue reprised interest in the North South Corridor being built by India, Russia and Iran as an alternative. https://www.businessinsider.in/business/news

Ships' energy efficiency management: organizational and economic aspect

Victoria Glavatskhih[1*], *Alexander* Lapkin[2], *Larisa* Dmitrieva[1], *Inna* Khodikova[1], and *Alexander* Golovin[1]

[1]Odessa National Maritime University, Department of Logistic and Project Management, 65029 Odessa, Ukraine
[2]Odessa National Maritime University, Department of Management and Marketing, 65029 Odessa, Ukraine

Abstract. Maritime transport plays an important role in economic development of society, because it carries about 85% of goods, but it consumes a significant amount of fuel resources, which has a serious impact on air pollution. Millions of tons of nitrogen oxides, sulfur oxides, soot, heavy metal compounds and water vapor are released into atmosphere each year. These factors influence constant search for more effective methods to reduce toxicity of exhaust gases and develop organizational and economic conditions for fuel usage.

1 Introduction

Despite the fact that maritime transport compares favourably with other types of transport in regards to profitability, it is also the first in regards to harmful gases emissions into atmosphere and environmental pollution. Marine vessels use heavy fuel, which contains a significant number of heavy and dangerous metals, and it has a significant impact on air and coastline pollution.

Annex VI (Rules for Air Pollution Prevention from Ships) to the International Convention for the Prevention of Pollution from Ships (MARPOL) 73/78 was implemented on 19 May 2005. Two sections are of the greatest importance for practice: Regulation 13, which establishes limits on nitrogen oxides (NO) emissions from marine diesel engines, and Regulation 14, which establishes limits on the sulfur content of marine fuels [1]. In October 2008, the 58th session of the International Maritime Organization (IMO) adopted a new version of Annex VI [2]. The new version of the rules establishes gradual (Tier I - Tier III) introduction of new standards with increasing requirements for them.

Harmful substances from ships that pollute marine environment include:

- fuel oil, lube oils and sludges;
- ballast water with certain chemical and biological content;
- toxic, anti-fouling coating.

Harmful substances that pollute atmosphere include:

- exhaust gases of ship power engines;

* Corresponding author: VGlavatskhih@gmail.com

- emissions of ozone-depleting and other gases contained in ship mechanisms and devices [3].

At the beginning of 2015, amendments to the Convention came into force, which regulate sulfur amount in marine fuel used on ships whose routes pass in SECA areas (trading areas, which include the Baltic, North Seas and the English Channel, where ships must switch to low sulfur fuel). Mandatory requirements were adopted by the IMO Committee for the Protection of Marine Environment in 2016 by amending MARPOL Chapter 4 of Annex VI to the Regulations for the Prevention of Air Pollution from Ships [4].

According to the new Regulation 22A on collection and submission of data about actual consumption of marine fuel [5], vessels with a gross tonnage of 5,000 tons and more are required to collect data on the consumption of each type of fuel they use, as well as other additional data, including transport indicators. activities. These vessels account for approximately 85% of carbon emissions from international shipping.

From January 1, 2020, the requirements of the MARPOL Convention to prohibit the use of fuels with a sulfur content of more than 0.5% are extended throughout the whole international trading area [6].

2 Results

Fuel costs are the main cost item when using maritime transport. Therefore, it is important to develop a set of measures aimed at reducing fuel consumption in ships operation.

On the one hand, shipping time is one of the most important indicators of shipping company quality. But on the other hand, vessels operation at optimal speeds allows to achieve a significant reduction in fuel consumption and, as a consequence, a reduction in operating costs and emissions from ships. Therefore, when choosing the optimal vessels speed, it is important to maintain a balance between these two indicators.

River and especially sea vessels travel long distances with a set, usually constant speed, at which engines run in optimal mode for a long time, and therefore the exhaust gases contain a minimum of toxic substances.

Under normal conditions of operation, the main sources of pollution are ship power engines, as well as ballast water and water used for washing cargo tanks. First of all, ship power engines pollute atmosphere with exhaust gases, and then toxic substances partially or almost completely get into waters of rivers, seas and oceans accordingly, thus volumes polluting land and water environment are statistically connected. Using maritime transport data, it is possible to estimate amount of harmful emissions and damage to the environment. Maritime transport carries a large number of oil cargoes, which increases effect of water environment pollution due to significant hydrocarbons weathering. At the same time, tankers account for more than 50% of sea pollution with oil and oil products from maritime transport.

CO, CO2, CH gases are heavier than air and are accumulated on the surface of water environment. CO and gaseous hydrocarbon emissions from vessels power engines are involved in oxidative reactions and eventually converted to CO2, which presence in atmosphere causes a greenhouse effect [7].

Considering that ship engines are the main consumers of high-sulfur heavy fuels - High-Sulphur Fuel Oil (HSFO), to regulate emissions of SOx, in 2015 control area were introduced by IMO - Sulfur Emission Control Areas (SECA), which are shown in Fig. 1.

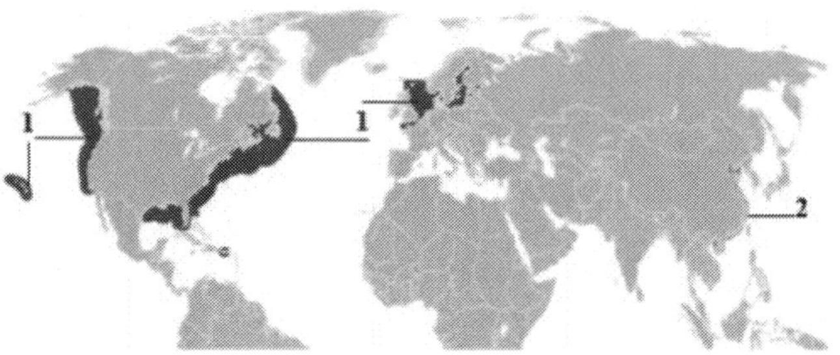

Fig. 1. SECA location and their limits: 1 - 0.1%; 2 - 0.5%.

The IMO, following allowability of using compatible low-oil fuel oil, has decided to globally limit the sulfur content of 0.5% for all trading areas, which came into force in 2020. This requirement is introduced in addition to the 0.1% sulfur limit in the North American and American areas of the Caribbean, North and Baltic Seas. In this case, vessels whose power engines are equipped with exhaust gas purification systems from sulfur content, will be allowed not to stop using HSFO (Table 1) [8].

Table 1. Regional and global sulfur regulators.

Region	Sulfur content limit in fuel	Possibility to use scrubbers
All trading areas in international waters	0,5 %	Yes
Control Areas (SECA)	0,1 %	Yes
The European Union	0,1 % in all port waters	In some countries, the use of open cycle is limited
China	0,5 % в in national waters in the 12-mile area	Yes
California	0,1 % in the 24-mile coastline area	No, only as an exception in the case of a separate application and additional research

Essential amendment to the regulation is "Carriage ban on HSFO in the Fuel System", with exception for vessels equipped with scrubbers. As earlier, it is allowed to carry HSFO as cargo, but it is forbidden to have HSFO in fuel system when scrubbers are not used. This amendment allows vessels using unsuitable fuel to be detained during Port State Control (PSC) without necessity to verify whether or not it has been used. It is expected that such amendments will significantly reduce possibility of violations of environmental legislation in international waters.

Expected increase in fuel costs, as many analysts of this sector predict, will lead to a certain slowdown in operating costs reduction. For the same reason, efforts will be focused on measures aimed at improving energy efficiency. Energy efficient vessels will be considered more competitive.

Vessels equipped with scrubbers can gain a significant competitive advantage. It is expected that vessels with scrubbers will initially be able to provide premium freight rates. However, if most vessels in a particular segment install scrubbers, the rates will be reduced as a result. Those vessels that will not be equipped with scrubber technology will be forced to reduce their rates to volatile levels and, ultimately, they may be pushed out of the market.

Therefore, it is very important for shipowners to monitor competition in their segment to ensure their dominance in the transport services market.

Significant investments necessary for competitiveness in the context of global sulfur oxide limits, combined with the costs associated with ballast water treatment systems, may lead to the situation when older vessels will become unprofitable for further operation, which may be the reason for higher disposal activity and then fleet renewal using modern, more efficient vessels. This trend may also accelerate introduction of alternative fuels, such as LNG [9].

Switching to distillate or MGO (Marine Gas Oil) fuel will mean a significant increase in fuel and lube oil costs, and may require modernization of power engines in terms of fuel preparation due to significantly low fuel viscosity. Fuel tanks previously used for HSFO must be thoroughly cleaned prior to MGO bunkering to avoid contamination problems and, as a result, non-compliance.

The main problem with the use of MGOs or distillates is related to their availability and cost in the fuel market. Many analysts believe [10] that in the first few months after new requirements introduction, the price difference between HSFO and distillates will be very high, which will significantly increase fuel cost and make alternatives financially attractive.

It is expected that low sulfur fuel blends will be available in the market as a variety of products. However, desulfurization engines are very expensive, and their installation, implementation and commissioning can take several years before they start operating at full capacity [11]. Thus, most refineries prefer to process higher grade fuels rather than invest in desulfurization systems. It can be assumed that to meet the demand will make new fuel mixtures that will meet the desired limit of sulfur content less than 0.5%, while reducing the cost by 10-15% compared to direct distillate fuels. The first fuel samples were presented at the end of 2018. It is likely that using new fuel mixtures will cause compatibility problems, and this will make fuel technological preparation a very important process for engine safe operation. Other problems associated with this type of fuel include long-term stability, impact on catalysts and relatively low flash point. Of great importance becomes quality control during bunkering, necessary to be confident that received fuel is in accordance with specification [12].

It is clear that using LNG as the main fuel will gain new benefits in context of new IMO limitations on sulfur emissions. LNG as marine fuel is technically proved solution, and bunkering infrastructure is developing rapidly around the world. Traditional fuel oils will remain the main fuel for most existing vessels in the near future, but commercial LNG opportunities are promising mainly for new ship designs, but in some cases also for conversion and modernization projects.

Thus, engine transferring to LNG should be carried out only on the basis of careful analysis of various information. In addition to commercial aspects, the main argument in favor of choosing LNG as a replacement for conventional fuel oils is a significant reduction in total air pollution from emissions of not only SOx but also NOx, as well as emissions of solids (Particulate Matter - PM).

Complete elimination of SOx and PM emissions – is potential to reduce NOx emissions by up to 85% - which contributes to using LNG, especially in control areas. In addition, LNG can reduce greenhouse gas emissions by 10-20% depending on engine technology. To sum up, using LNG has many benefits for both human health and the environment. addition, it also has a positive effect on ship's Energy Efficiency Design Index (EEDI). [13].

Initial data for calculating emissions of harmful substances and their dispersion into atmosphere are the annual consumption of fuel and materials consumed by emission sources, including merchant navy vessels [14].

The maximum emission of i-th substance (g/s) from diesel engine is determined by formula:

$$Mi = (1/3600) \cdot eMi \cdot Pe \quad , \tag{1}$$

where eMi – emission of i-th harmful substance per unit of useful diesel engine operation rated power mode, g/kWh; Pe – diesel engine operating power, whose value is taken from manufacturer's technical documentation, kW.

Under unstable economic situation in the world, maritime industry is facing changing operating conditions. In order to remain competitive in the market, most shipping companies are forced to take measures to improve energy efficiency. This is partly due to changes in oil and fuel prices. Therefore, a lot of attention of shipping companies is focused on developing measures to reduce fuel costs. These include the following: measures to organize transportation (vessels redistribution; routes integration; operating speed reduction; service cessation in less profitable areas; improvement of navigation conditions), measures for technological improvement (improvement of hull design; new technologies for marine engines; using alternative energy sources; computer technology).

Next, let us consider one of the most effective measures to reduce fuel costs, which lies in vessel operating speed reduction. To do this, we will analyze dependence of fuel consumption on vessel speed. Isuues on dependence of fuel consumption on vessel speed and methods of choosing optimal operating speed was studied by many scientists, including [15-17].

Fuel savings during vessel operation might be achieved in several ways. One of them is to renew fleet with perfect vessels of new generation, which are equipped with modern economical engines. This way involves attracting significant investment. Another way to save fuel during vessel operation is to select and ensure optimal modes of marine engines operation, as well as to carry out various organizational and technical measures to save fuel and energy resources. Reducing speed during vessel operation allows the shipowner to reduce vessel operating costs, which is widely used in practice.

Based on known dependence of fuel consumption per day underway ($q_{underway}$) on vessel speed (V):

$$q_{underway} = \left(\frac{V}{V_{max}}\right)^a * (q_{max} - q_{min}) + q_{min}, \quad (4)$$

where a - constant parameter (a=2,507849 is set using the least squares method);
V_{max} - maximum vessel speed, knots;
q_{max}- fuel consumption at maximum speed, t / day;
q_{min}- fuel consumption at minimum speed, t / day,
it is possible to determine the most appropriate vessels speed modes.

We have made calculations of fuel oil and diesel fuel consumption dependence by bulk carriers underway (type Handysize) with deadweight ranging from 32 thousand tons to 43 thousand tons.

As it can be seen from Figure 2 and Figure 3, at speed range up to 12 knots, the lowest fuel consumption refers to vessel "D", however, at range of more than 12 knots – priority is taken by vessel "A". And diesel fuel consumption varies in the range: from 12 knots for vessel "C" to 14.5 knots for vessel "A".

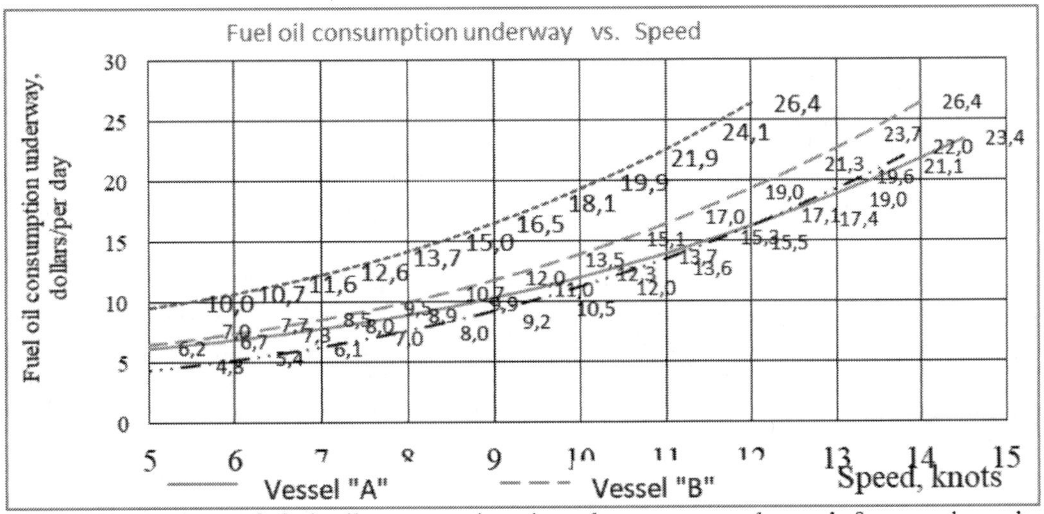

Fig. 2. Comparison of fuel oil consumption dependence on vessel speed for vessels under consideration.

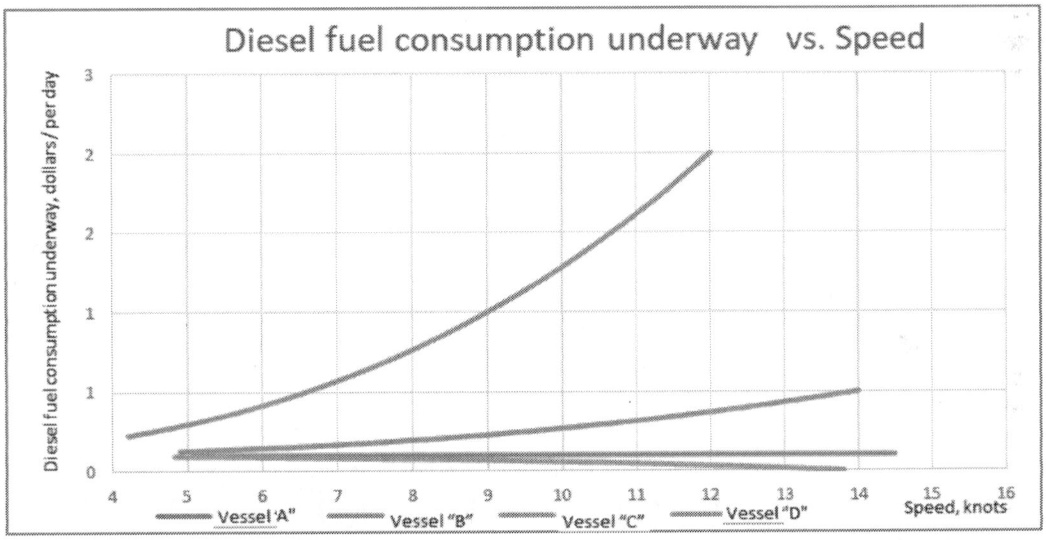

Fig. 3. Comparison of diesel fuel consumption dependence underway for vessels under consideration.

When choosing a vessel, you should also pay attention to the length of speed range at which vessel is optimally used. If vessel usage is optimal in a wide range of speeds, as in our case, for vessels "A", "C" and "D", shipping company in the future can provide additional flexibility in agreeing freight rates and shipping time through possibility of vessel efficient operation at different speeds, which also affects amount of harmful emissions into atmosphere and coastline areas [14].

3 Conclusions

It is important for maritime transport operation to reduce unit costs, the most important component of which is fuel costs and energy efficiency. Factors that determine rationalization process for their use can be divided into certain groups:

1. Technological factors of resources efficiency include measures related to improving transport production. The same group of factors includes increasing level of mechanization and automation, rational consumption of fuel oil and lube oil, reducing norms of their usage on the basis of modern experience.

2. Constructive economy factors include group of measures aimed at improving design solutions and calculation methods. Especially important for marine transport is to increase operational reliability, reduce mass and dimensions of engineering factors.

3. Organizational and economic factors include measures aimed at improving organizational and economic conditions of using fuel. The basis of these measures lies in improving norm setting and efficiency of fuel and energy resources, improving conditions of their storage, utilization of thermal energy.

To sum up, the basis for fuel and energy efficiency in maritime transport is development and implementation of energy-saving and environmental technologies and improving fuel management.

References

1. Semanov, G.N. (2013). Harmful Air Emissions from Ships: Towards IMO Standards. SPb.: Publishing house Science and transport. Sea and river transport.
2. MARPOL (2019). Revised MARPOL Annex VI. Rules for the Prevention of Air Pollution from Ships. http://rise.odessa.ua/texts/MARPOL_VI_19.php3
3. Bellona (2007). International Management Code for the Safe Operation of Ships and Pollution Prevention - ISM Code (Chapter IX SOLAS-74) ISM CODE. https://bellona.ru/2007/05/08/mezhdunarodnyj-kodeks-po-upravleniyu-b
4. MARPOL (2018). Guidelines for the application of the provisions of the International Convention MARPOL 73/78 (revised in 2018). https://files.stroyinf.ru/Index2/1/4293738/4293738163.htm
5. Verkhovna Rada of Ukraine (). About the consolidated Provisions about collecting the tributes, which are used in the form of the given tributes. https://zakon.rada.gov.ua/laws/show/835-2015-%D0%BF#Text
6. IMO (2020). Prospects for the shipping industry and refineries. https://modcon.ru/applications
7. European Environment Agency (2020). Atmospheric greenhouse gas concentrations. https://www.eea.europa.eu/data-and-maps/indicators/atmospheric-greenhouse-gas-concentrations-6/assessment
8. DNV GL (2021). Global Sulfur Cap 2020 - extended and updated. https://www.dnvgl.com/maritime/publications/global-sulfur-cap-2020.html
9. Petrov, A. P., & Zhivljuk, G.E. (2019). Ecological safety of ship internal combustion engines. Selecting the efficient method for compliance with the new requirements of 2020 for sulfur emissions. Vestnik Gosudarstvennogo universiteta morskogo i rechnogo flota imeni admirala S.O. Makarova, 727–744.
10. Timofeev, O. Ya., Valdman, N.A., & Kryzhevich, M.I. (2014). Features of the introduction of new requirements for the environmental safety of ships and offshore installations in the northern seas and on the Arctic shelf. Arctic: ecology and economics, 3 (15), 79-85.
11. Leontyev, A. I. (2018). Cleaning of diesel fuel. Desulfurization by the adsorption method. Business magazine Neftegaz.RU, 8, 62–70.
12. Petrov, A. P., & Zhivljuk, G.E. (2019). Environmental safety. Limitation of sulfur emissions by the ship power plants. Vestnik Gosudarstvennogo universiteta morskogo i rechnogo flota imeni admirala S. O. Makarova, 130–145.

13. Ivanchenko, A. A., Petrov, A. P., & Zhivljuk, G.E. (2015). Energy efficiency of ships and regulation of greenhouse gas emissions. Bulletin of the State University of Marine and River Fleet named after Admiral S. O. Makarov, 3 (31), 103–112.
14. Lapkina, I.A., & Glavatskikh, V.I. (2020). Management of the exploitation of vitrates of ships and the rules of MARPOL. II International Scientific and Practical Maritime Conference of the Department of SEU and TE of Odessa National Maritime University, MPP & O-2020 (Marine Power Plants and Operation), 376-381.
15. Lapkina, I.O., Malaksiano, M.O., & Glavatskykh, V.I. (2019). To the issue of the possibility of operating vessels at slow speeds Vernadsky, 30 (69), 134-140.
16. VG Marine Service (n.d.). Ecology of ship systems and crew. http://vgms.com.ua/esse-ecology-ship-systems-equip
17. Lapkina, I.A., & Akimova, O.V. (2011). Determination of the optimal operating speed of container ships when changing the volume of traffic on the line. Method for managing the development of transport systems, 18, 165-181.

Financial and investment processes in ensuring the development of marine infrastructure

Larysa Hryshyna[1*], *Elena* Pogoryelova[1], *Petro* Karas[1], *Natalia* Mihai[1], and *Irina* Khmarska[1]

[1]Admiral Makarov National University of Shipbuilding, Prospekt Heroyiv Ukrayiny, 9, 54007, Mykolayiv, Ukraine

Abstract. The article identifies the place and importance of financial and investment processes in ensuring the development of maritime infrastructure as a component of the maritime complex of Ukraine. The current state of maritime infrastructure and the level of its use in the transformational conditions of the domestic economy are studied. Factors influencing the functioning and development of marine infrastructure are identified. The main directions of intensification of financial and investment processes to ensure the effective functioning and development of the maritime industry and its infrastructure are identified.

1 Introduction

One of the strategic directions of sustainable development of Ukraine and its integration into the Euro-Atlantic space is the formation of the country as a maritime state.To implement them, it is necessary to solve a number of problems in the field of maritime activities, including: weakening the competitive position of Ukrainian seaports in the international and regional market of port services, reducing business activity in shipbuilding and ship repair, exclusion of Ukraine from transit cargo flows and loss of transit potential; inconsistency of the state of marine infrastructure with the needs of the national economy. As the experience of other countries shows, the most attractive components of the transport system are water (sea) transportation, as the most efficient way to deliver goods due to its low cost and ability to solve problems of transportation of large cargo. The current state of maritime infrastructure shows its significant moral and physical performance, and financing its renovation and development has remained an unresolved issue for many years.

2 Literature review

Problems of functioning of the enterprises of the sea branch, formation and development of the sea infrastructure are considered in works of domestic scientists and practitioners, among which it is necessary to note such as G. Yefimova [1], I. Irtyshcheva [2],

*Corresponding author: grishinappi@gmail.com

Y. Makogon [3], N. Prykhodko [4], M. Primachev [5], V. Vinnikov [5], A. Vorobyova [6], V. Parsyak [6], M. Fateev [6]. Most of the scientific works devoted to the study of the problems of restoration and development of the maritime industry are considered in the context of recognizing Ukraine as a maritime state with a long coastline, potentially developed maritime complex in ensuring transportation of goods through domestic and transit ports. However, a number of pressing issues of financial and investment support for the development of marine infrastructure remain, which need further research.

The purpose of the article is to study the financial and investment processes as a prerequisite for ensuring the development of maritime infrastructure in modern conditions of the domestic economy.

3 Results

One of the important components of the formation and further development of Ukraine as a maritime state is the adoption of the Maritime Doctrine of Ukraine for the period up to 2035 (entered into force on December 18, 2018), which aims to improve the efficiency of all marine resources and successful implementation of state maritime policy. which depend on the sovereignty and territorial integrity of Ukraine, its economic and energy independence, sustainable development and integration into the Euro-Atlantic space [7].

It should be noted that as part of the maritime complex and transport system, maritime infrastructure connects all regions and continents, which is an integral part of ensuring the integrity and unity of the world economic space. Therefore, the very concept of "marine infrastructure" has a wide range of understanding and is used to describe collective engineering structures and facilities, buildings, systems and services located within the waters to ensure effective maritime business.

Maritime infrastructure includes ports, sea lanes, docks, berths, warehouses and other facilities of entities engaged in various cargo operations. The value of these objects increases in the context of globalization of economic interests and strengthening of integration processes, when maritime transport is an important factor in the formation of world economic cooperation in maritime activities. Globalization can be seen as a process of growth of the unit cost of tonnage in a single economic space. Limited resources, distribution of products and goods depends on the level of efficiency of transport opportunities. The dependence of the development of the world economy on the technical and economic level of water transport determines the task of a balanced system of formation of its production and financial results [5].

According to the results of the analytical study of the current state of use of maritime infrastructure in Ukraine, there is a low level of use of the potential of the maritime industry and its infrastructure (Table 1) [8, 9]. In particular, if in 2018, compared to 2010, the freight turnover of all transport decreased by 13.7%, the transportation of goods by sea has halved. Even more negative dynamics of export-import services is observed in maritime transport.

If in 2010 the share of exports of services provided by sea transport was 15.6% of their total, in 2016 they amounted to 8.0%, in 2017 - 10.4%, in 2018 - 8.9%.

The level of capital investment directed to the development of maritime transport has a steady trend towards minimal investment. If in 2018 the total amount of capital investment in transport increased compared to the previous period by UAH 12.2 billion (32.2%), in maritime transport, on the contrary, there is a reduction of UAH 1.1 billion (38%).Thus, such trends in the domestic maritime sector, general unfavorable economic and political conditions cause a gradual loss of Ukraine's potential in servicing export-import maritime cargo flows, and due to the high level of wear and tear of the maritime sector accelerates the competitiveness of domestic maritime enterprises.

Table 1. Dynamics of indicators of activity of sea and water transport of Ukraine.

Indicator	Years									
	2010		2016		2017		2018		2019	
	amount	(%)	amount	(%)	amount	(%)	amount	(%)	amount	(%)
Freight turnover, total, billion.tkm, incl. -maritime transport -river transport	418,7 3,6 1,7	100 1,2 0,9	344,2 2,5 1,5	100 0,7 0,4	364,2 2,9 1,4	100 0,8 0,4	361,3 1,8 1,6	100 0,5 0,4	355,0 1,8 1,6	100 0,5 0,4
Cargo turnover of maritime transport by types of communication, total, billion tkm, incl. -foreign -shoot	5,17 4,92 0,25	100 95,2 4,8	2,54 2,22 0,32	100 87,4 12,6	2,85 2,38 0,47	100 83,5 16,5	1,82 1,58 0,24	100 86,8 13,2	78,0 22,0 1,77	100 1,38 0,39
Capital investmens billion UAH -including water transport	19,6 0,18	100 0,9	25,1 0,24	100 1,0	37,9 0,25	100 0,7	50,1 0,2	100 0,4	43,8 0,3	100 0,7
Average number of full-time employees, thousand people -including water transport	824,3 9,6	100 1,2	659,9 3,4	100 0,5	655,2 2,3	100 0,4	648,4 2,6	100 0,4	635,1 2,5	100 0,4
Export-import of transport services, Exports, million US dollars -of them maritime transport services Imports, million USD -of them maritime transport services	7835,2 1 234,3 1178,9 143,0	100 15,6 100 12,2	8300,5 661,6 989,2 141,2	100 8,0 100 14,3	5861,4 612,1 1213,1 222,8	100 10,4 100 18,4	5851,4 523,3 1464,8 223,0	100 8,9 100 15,2	9109,9 595,5 1559,1 273,8	100 5,87 100 17,6

Source: Compiled on the basis of statistical data [8, 9]

The functioning and development of maritime infrastructure occurs under the influence of many factors that are both stimulating and deterrent or destructive (Fig. 1). In our opinion, among these factors, one of the most important should be noted financial and economic, as well as the quality of state support for the development of the maritime industry as a whole. Therefore, the financial and investment process should be considered the basis for the future development of maritime infrastructure, intensification of basic and applied research in domestic shipbuilding, restoring the competitive potential of the maritime transport sector of Ukraine as a whole.

Businesses operating in the seaport need to attract additional financial resources (for dredging, construction of modern shore and raid terminals, upgrade of transhipment equipment, port infrastructure) in the amount of more than 3 billion US dollars, which leads to favorable conditions for efficient investing in the economy.

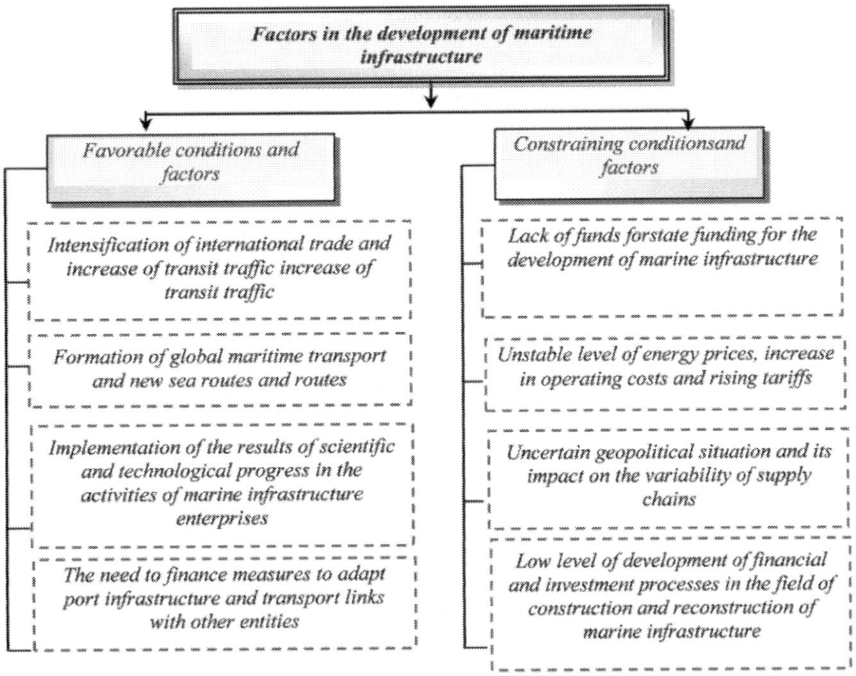

Fig.1. Factors and conditions of marine infrastructure development [1, 2, 4, 6].

According to the activities of the Administration of Sea Ports of Ukraine (AMPU), in the first half of 2019 increased capital investment in the development of port infrastructure, funding for projects to maintain passport depths in domestic ports and fleet repairs increased 1.6 times. According to the head of the AMPU, the unstable dividend policy of the state complicates the mechanism of long-term planning for the development of the port sub-sector and does not contribute to the implementation of projects for the development of maritime (port) infrastructure in the coming years.

In total, during the period of its activity AMPU has invested more than UAH 6,6 billion in the development of the port economy, and in 2020 plans to increase its profit to UAH 6,7 billion [10]. Great hopes are also placed on the activity of the Maritime Administration, the main tasks of which are: implementation of state policy in the field of sea and river transport, merchant shipping, navigation and hydrographic navigation, state control over safety, as well as formation of financing mechanism for maritime industry and its infrastructure software. But during the two years of its operation, funding issues remain unresolved, although work has been done on participation in international state donor platforms, support for the European transport program in the formation of the material and technical base of the industry.

An important event in the economic space of Ukraine was the holding in October 2019 of the First Investment Forum "RE: THINK invest in Ukraine" under the auspices of the President of Ukraine Volodymyr Zelensky [11].

Among the 120 investment projects should be noted 2 projects for the development of maritime infrastructure of the Black Sea region, the main indicators of which are presented in table.2.

Table 2. Business opportunities for financing investment projects for marine infrastructure development.

Specialized seaport "Olbia" **Business opportunity: Seaport concession**	**Kherson sea trade port** **Business opportunity: Seaport concession**
Estimated investment amount - $ 49 million USA	*Estimated investment amount -* $ 45 million USA
Form of ownership: Public-private partnership Location: Mykolayiv The essence of the project: • Transfer of state property to a private partner-concessionaire for a period of 35 years • Carrying out activities in the field of construction and operation of seaports • A prerequisite for the project is the construction by the concessionaire of a grain terminal for 2 million tons per year	*Form ownership*: Public-private partnership Location: Kherson The essence of the project: • Transfer of state property to a private partner-concessionaire in order to modernize the impaired assets of the port • Transfer of berths and in-port facilities to a private investor in accordance with the concession agreement • Potential traffic - up to 4 million tons annually until 2030
Area: 145 hectares Existing buildings: • 6 berths, 57 buildings and structures, including 2 general cargo depots, 20 open storages Existing equipment: • 22 gantry cranes, 145 units of handling equipment, 3 tugs	Area: 195 hectares Term of concession: 30 years Financial indicators Initial investment: more than 10 million US dollars Concession payments: permanent (340 thousand US dollars) and variables (over 0.73% of annual profit) Reimbursement of 1.3 million euros for the development of a feasibility study
Implementation period: depending on the business project	*Implementation period*: depending on the business project
Socio-economic effect Pilot project of the concession under the new transparent rules Restoration of the seaport and increase of grain exports Increasing investment attractiveness of the city of Mykolayiv	*Socio-economic effect* Pilot project of the concession under the new transparent rules Restoration of the seaport and increase of grain exports Increasing the investment attractiveness of the Kherson region

Source: Compiled on the basis of data [11]

Given the current trends of the economic system of Ukraine, it is the development of maritime infrastructure will contribute to the realization of the potential of the agro-industrial complex, the scope of which covers almost three quarters of the country and generates more than 10% of GDP [8]. Compared to other industries, it is the agricultural sector that shows positive dynamics, occupies a significant share in agricultural exports and can ensure the preservation of competitive positions in the world food market.

In addition, the ongoing processes of decentralization, the creation of new communities, which received from the state more than 2 million hectares of agricultural land provide unique opportunities for the development of new technologies for production that can compete in the European and world food market. In the conditions of transformation of land ownership relations in Ukraine it is possible to increase long-term investments, lending to the agricultural sector of the domestic economy.

This, in turn, will contribute to the development of coastal regions, increase employment and ensure the socio-economic level of development. A positive example is the creation in the Black Sea region of production terminals for the sale of agricultural products, in particular, grain and oilseeds. Evidence of this is the activity of the company "Nibulon", which in July 2020 shipped for export on an external raid of the Mykolayiv seaport without ships in Mykolayiv 387,28 thousand tons of grain, almost 453 thousand tons of cargoes were transported by rivers (Nibulon official website). Effective foreign economic activity and a stable image of the company have made it a reliable partner for cooperation within the EU and abroad.

It is important to note that today Nibulon is one of the largest investors in Ukraine's economy - more than 2,3 billion dollars. US - and the undisputed leader in the revival of national shipbuilding - $ 600 million USA [12]. In recent years, the company has headed the top 3 shipyards in the ranking of the Association of Shipbuilders of Ukraine "Ukrsudprom" as the most productive shipbuilding company[9]. Today, the shipbuilding and ship repair plant "Nibulon" is a completely modern enterprise, which is almost the only one in Ukraine that builds complete ships. It specializes in the construction of ships and vessels up to 140 meters long. The company has made a significant contribution to the reform of marine infrastructure, modernized the factory slip and embankment, renovated the crane industry, purchased the best machines and welding machines, built new shops and equipped them with state-of-the-art equipment from the world's leading manufacturers. Nibulon's fleet consists of 82 vessels for various purposes and is the newest and most modern in Ukraine. And the company's compliance with the strict requirements of international ISO standards is a confirmation of the sustainable development of the company and a significant contribution to the global market of shipbuilding and ship repair, development of marine infrastructure. One of the levers of regulation of financial and investment processes in the maritime sector is the creation of favorable investment conditions for the development of seaports and large-scale investment projects, including by improving the Laws of Ukraine "On Concessions", "On Seaports of Ukraine", "On Fisheries" , industrial fishing and protection of aquatic bioresources, adoption of the procedure and conditions for compensation of investments involved in the development of strategic objects of port infrastructure, simplification of procedures for transfer of property and land located in the port in concession; implementation of the relevant depreciation policy.

Note that in today's difficult post-crisis economic conditions (inflation, rising steel prices, lack of financial resources, rising energy prices and components) the activities of domestic shipbuilders have not yet become the key to the formation of competitive shipbuilding and ship repair as important components of Ukraine's maritime complex [4].Therefore, it is important to form and implement an effective financial and credit mechanism to support expanded reproduction in shipbuilding by providing state guarantees for loans (loans) attracted for investment projects; full or partial compensation at the expense of the state or local budgets of interest on loans obtained by business entities for the implementation of investment projects in shipbuilding; implementation of investment projects at the expense of loans (borrowings) attracted by the state on the basis of international agreements of Ukraine; state financing of leasing companies that order vessels from domestic shipbuilders.

It should be noted that an important aspect of the formation of financial and investment processes in ensuring the development of maritime infrastructure is the relationship of the maritime sector with financial institutions.

The main requirements should be large amounts of borrowed capital, long terms and the availability of its price. The sources of such funds may be funds of legal entities and individuals, funds of public authorities, as well as the capital of banking institutions. If it is not possible to access alternative financing, then maritime infrastructure entities can use leasing services, which is quite common in developed maritime countries.

As the positive experience of other countries shows, it is advisable to create banking consortia, which are headed by a bank with sufficient liabilities, extensive lending experience and which has well-established links with financial and credit institutions. The operation of such a banking consortium can have many benefits for the development of maritime infrastructure [6].

We should not forget about the financing of innovation, the importance of which for the development of the maritime industry is difficult to overestimate. But, unfortunately, the level of financial support for the introduction of new technologies in shipbuilding, the use of innovative processes in the activities of maritime infrastructure remains low. It is the lack of equity in the enterprises of the maritime industry, which is the main source of costs for innovation, is the main reason for the low innovative level of products and services of the maritime industry [13].

For the adoption of scientifically grounded decisions on the directions of activation of financial and investment processes to ensure the effective functioning and development of the marine industry and its infrastructure it is advisable to use economic and mathematical models, which can be quantitatively described by the connection between the input parameters of the economic system and the productive indicator. In this study, a method of correlation-regression analysis is used, which includes the determination of the object sunder investigation, analysis of interconnections between individual variables, the choice of rational type of econometric model.

To build a model for the formation of strategic directions of financial and investment processes in ensuring the development of marine infrastructure, data are formed in the form of a spatial and time sampling. In particular, these data are determined as a result of the examination of the main indicators of the marine industry for 2010-2019 using the system of indicators: volumes of transported cargoes in the marine industry (X1), exports of maritime transport services (X2), import of maritime transport services (X3). In order to find out the impact of each of them, the econometric model between the misbuilt on the volume of financing of marine infrastructure:

$$Y = f(X1; X2; X3; u), \quad (1)$$

where u – is a stochastic component of the model.

Based on empirical data, the values of regression coefficients were evaluated, which enabled to choose a linear model presented by such a function:

$$\hat{y} = -261{,}81 + 0{,}2075X1 + 0{,}5032X2 + 1{,}3314X3 \quad (2)$$

It is worth noting that the calculated values of the determination coefficient R^2 =0,858, Fischer's criterion $F_{розр}$=7,18 ($F_{табл}$ = 2,69) and the adjusted coefficient of 0,743 indicates a sufficient compliance with the empirical data of the proposed econometric model. Thus, this confirms the expediency of using the proposed model in the formation of strategic directions of intensification of financial and investment support for the development of marine infrastructure.

Regarding the financing of maritime infrastructure development through investments of foreign capital, it should be noted that they can be carried out in various forms, and the most common form is investment in share capital of enterprises, government and corporate securities.

One of the largest partners in such cooperation is the European Bank for Reconstruction and Development (EBRD), which provides international loans on concessional terms under state guarantees and lends to private enterprises on commercial terms. Among the goals and areas of financial support are measures to complete the modernization of the main transport corridor connecting Ukraine with the European Union; project initiatives in seaports, private terminals and infrastructure. As of October 1, 2020, the total amount of EBRD funding is € 14,739 billion under 479 projects. 8 joint projects with the EBRD with a total loan amount of EUR 1,701.0 million are under implementation, of which only EUR 912,58 million was spent (53,65% of the total loan amount [14]. But due to a number of global changes, political and economic events in Ukraine, the domestic maritime industry does not have to rely on the help of foreign investors, unfortunately. In particular, in the Nikolaev area, as one of the leading regions of development of sea branch, decline of investment activity is observed. However, none of these projects provides financial support for the implementation of important initiatives for the domestic economy in the development of maritime activities.

4 Conclusions

Modern post-crisis conditions of the domestic economy dictate the need to intensify the use of strategic resources and potential of the maritime industry and its infrastructure, which actualizes the solution of financial and investment processes in the maritime sector. Among the main areas of solving the problem of increasing financial resources in ensuring the development of the maritime industry and its infrastructure should be identified:

– revival of the competitive Ukrainian merchant fleet by providing legislative support to Ukrainian shipping companies, application of economically justified tax and customs regimes of shipping and shipbuilding, in particular tax and customs regimes;

– participation in the implementation of TRACECA projects related to transport infrastructure in the Black Sea region;

– introduction of a transparent system of financing activities for the implementation of Ukraine's international obligations in the field of merchant shipping;

– establishment of transparent and economically sound methods of calculation, procedure for collection and exclusively targeted use of port dues and improvement of tariff policy in the port industry;

– development of the internal market of shipbuilding with the use of mechanisms of public-private partnership, implementation of the state order for the production of certain types of shipbuilding products;

– development of regional investment maps of coastal regions in the areas of «marine industry» and «recreational and tourist sphere».

Thus, further intensification of financial and investment processes in ensuring the development of maritime infrastructure requires the formation of an appropriate organizational and economic mechanism, the introduction of which will enhance investment activities, optimize sources of funding for the maritime sector, innovative development of port infrastructure and more.

To intensify the investment market, the state's credit policy must also be improved. At the same time, an important task of public policy should be to create favorable conditions for intensifying long-term lending, which is especially important for the maritime industry and will be possible through measures to increase lending by commercial banks, soften interest rate policy, reorient credit resources to long-term lending, creation of special investment banks with the provision of preferential terms for lending to investment programs to ensure the development of maritime infrastructure and support the formation of Ukraine as a maritime state.

References

1. Yefimova, G. (2015). Scientific principles of strategy of development of shipbuilding enterprises of Ukraine. Economist, 6, 17-24.
2. Irtyshcheva, I., Stehnei, M., & Gurina, O. (2018). Financial mechanism of the socio-oriented economic development of the black sea region. Baltic Journal of Economic Studies, 2018, 4 (4), 202-208. https://doi.org/10.30525/2256-0742/2018-4-4-202-208
3. Makogon, Yu. (2018). External economic potential of Ukraine in the black sea region. Economic Bulletin of NTUU "KPI", (15), 114-124. https://doi.org/10.20535/2307-5651.15.2018.135937
4. Hryshyna, L., Karas, P., & Prykhodko, N. (2017). Development of shipbuilding industry in the context of reforming the transport system of Ukraine. Bulletin of Transport Economics and Industry, 58, 86-96.
5. Miyusov, M., Primachev, N., & Vinnikov, V. (2006). Strategy of positioning of national sea transport in the global transport market. Odessa: Autograph.
6. Vorobyova, I., Yefimova, G., Karas, P., Kirichenko, O., Kramarenko, O., Parsyak, V., & Khachaturov, E. (2012). Finance of shipbuilding enterprises: monograph, Nikolaev: NUC.
7. Cabinet of Ministers of Ukraine (2018). Maritime Doctrine of Ukraine for the period up to 2035. As amended by the Resolution of the Cabinet of Ministers of Ukraine of December 18. 1108. https://zakon.rada.gov.ua
8. State Statistics Service of Ukraine (2020). http://www.ukrstat.gov.ua
9. State Statistics Service of Ukraine, (2019). Transport and Communications of Ukraine 2019. http://www.ukrstat.gov.ua
10. Seaports Administration of Ukraine (2020). http://www.uspa.gov.ua
11. European Union's EU4Business Initiative (n.d.). RE: think. Invest in Ukraine. The first investment forum in Ukraine. https://eu4business.eu/uk/events/mizhnarodnyy-investyciynyy-forum-rethink-invest-ukraine
12. Nibulon (n.d.). https://www.nibulon.com/
13. Association of Shipbuilders of Ukraine (n.d.).. http:// www.ukrsudo.kiev.ua
14. Ministry of Finance of Ukraine (2020). International cooperation: EBRD. https://www.minfin.gov.ua

AUTHOR INDEX

Abesadze Nino 51
Bakuradze Ivane 42
Baratashvili Nazi 32
Bogomolova Kateryna 148
Budz Olena 121
Chaganava Paata 15
Chitaladze Ketevan 51
Chkhaidze Irma 32
Devadze Anzor 15
Devadze Lela 15
Dmitrieva Larisa 176
Erkomaishvili Gulnaz 1
Filimonova Olena 73
Gechbaia Badri 1
Giorgobiani Maia 51
Glavatskhih Victoria 176
Golovin Alexander 176
Hordopolov Volodymyr 89
Hryshyna Larysa 112, 184
Hryshyna Nataliia 112
Iefimova Ganna 103
Ishchenko Natalia 73
Jabnidze Nato 32
Karas Petro 184
Katamadze Davit 42
Katamadze Guliko 42
Kharaishvili Eteri 1
Khmarska Irina 112, 184
Khodikova Inna 176
Kinkladze Rusudan 51
Kl³ukha Oksana 141
Kordzaia Irakli 15
Kovshun Nataliia 141
Kovtun Natalia 156
Kralia Viktoriia 148
Kupraty Oleksandr 128
Labartkava Andrei 103
Lapkin Alexander 176
Lee Jade Man-yin 163
Lobzhanidze Manana 1
Lopatin Artem 73
Makharadze Nanuli 32
Mamuladze Gela 42
Mamuladze Giorgi 42
Mayorova Irina 81
Mihai Natalia 112, 184
Milinchuk Olga 121
Mitkov Milen 23
Mysiuk Viktoriia 89
Natsvlishvili Ia 1
Nazarova Karina 89
Nezhyva Mariia 89
Nosovckaj Oksana 81
Ocheredco Elena 81
Pashchenko Oleksiy 103
Petrova Mariana 23
Podolska Olha 148
Pogoryelova Elena 112, 184
Radukanov Sergey 23
Robitashvili Natalia 51
Romaniv Oksana 121
Rudenko Natalia 73
Ryabchuk Valentyn 81
Sagareishvili Nino 65
Saghareishvili Mariam 58
Sahachko Yuliia 148
Savina Nataliia 141
Smihunova Olena 148
Temerbek Alisa 81
Tsetskhladze Leila 32
Ventsuryk Alisa 141
Wong Eugene Yin-cheung 163
Yarmolyk Dina 121
Yushchenko Nadiia 156
Zhemba Alla 141